A LIFE OF
Ernest Starling

American Physiological Society
People and Ideas Series

Circulation of the Blood: Men and Ideas
Edited by Alfred P. Fishman and Dickinson W. Richards
1982

Renal Physiology: People and Ideas
Edited by Carl W. Gorschalk, Robert W. Berliner,
and Gerhard H. Giebisch
1987

Endocrinology: People and Ideas
Edited by S.M. McCann
1988

Membrane Transport: People and Ideas
Edited by Daniel C. Tostcson
1989

August & Marie Krogh: Lives in Science
Bodil Schmidt-Neilsen
1995

Respiratory Physiology: People and Ideas
Edited by John B. West
1996

Moving Questions: A History of Membrane Transport and Bioenergetics
Joseph D. Robinson
1997

Exercise Physiology
Edited by Charles M. Tipton
2003

A Life of Ernest Starling
John Henderson
2005

A LIFE OF

Ernest Starling

JOHN HENDERSON

Published for the
American Physiological Society
by

OXFORD
UNIVERSITY PRESS

2005

OXFORD
UNIVERSITY PRESS

Oxford University Press, Inc., publishes works that further
Oxford University's objective of excellence
in research, scholarship, and education.

Oxford   New York
Auckland   Cape Town   Dar es Salaam   Delhi   Hong Kong   Karachi
Kuala Lumpur   Madrid   Melbourne   Mexico City   Nairobi
New Delhi   Shanghai   Taipei   Toronto

With offices in
Argentina   Austria   Brazil   Chile   Czech Republic   France   Greece
Guatemala   Hungary   Italy   Japan   Poland   Portugal   Singapore
South Korea   Switzerland   Thailand   Turkey   Ukraine   Vietnam

Published by Oxford University Press, Inc.
198 Madison Avenue, New York, New York 10016

www.oup.com

Oxford is a registered trademark of Oxford University Press

Library of Congress Cataloging-in-Publication Data
Henderson, John, 1934–
A life of Ernest Starling / John Henderson.
p. cm. — (People and ideas series)
Includes bibliographical references and index.
ISBN-13 978-0-19-517780-0
ISBN 0-19-517780-0
1. Starling, Ernest Henry, 1866-1927.   2. Physiologists—Great Britain—Biography.
I. Title.   II. Series.

QP26.S714H46 2005
612'.0092—dc22 [B]   2004054741

Printed and bound by CPI Group (UK) Ltd, Croydon, CR0 4YY

Transferred to Digital Print 2011

In memory of two teachers,
Sandy Ogston
and Peter Daniel
with gratitude and affection.

Preface

*When I compare our present knowledge of the workings of the
body, and our powers of interfering with and controlling these
workings for the benefit of humanity, with the ignorance and
despairing impotence of my student days, I feel that I have had
the good fortune to see the sun rise out of a darkened world . . .*
                                                    —Ernest Starling, 1923

Medicine leans heavily on understanding how the body works. The study of
the body's normal workings is physiology, and Ernest Starling was an exceptional physiologist who believed passionately that medicine could not advance without this contribution. As he was also a good physician, he was in
the best position to ask the best questions—ones that could be answered by
experiment. And this is how he spent a large part of his life.

Born in 1866 into a large Victorian family, Starling had an outstanding
career as a student and doctor at Guy's Hospital. Rejecting the obvious life of
a Harley Street physician, he launched himself into research and became the
first full-time physiologist at Guy's. In 1899 he was made Professor of Physiology
at University College, London (UCL) and worked there until his death in 1927,
aged 61. He was center-stage in a remarkable British flowering of physiology,
and contributed significantly to at least four separate areas of the subject:

1. The balance of hydrostatic and osmotic forces at the capillary ("Starling's
   Principle").
2. The discovery of the hormone secretin (along with his brother-in-law,
   William Bayliss) and the introduction of the word "hormone."
3. The analysis of the heart's activity as a pump ("Starling's Law of the
   Heart," "The Frank-Starling Law").
4. A number of fundamental observations on the action of the kidney.

These are the bare facts. Many of his contemporaries, including long-forgotten physicians and physiologists, were given knighthoods; Starling received nothing. The Nobel committee seemed to pass him by. The physiologist J.C. Eccles (a Nobel laureate) wrote, "This discovery [of secretin, the hormone] was recognized by the Nobel committee in 1913–1914 to deserve a prize, which surely would have been awarded but for the long period of suspension of awards during the 1914–18 war" (Eccles, 1971). Does the passage of four years wipe out a remarkable scientific achievement? Is Eccles' story true? There are clearly ways in which Starling's achievements didn't seem to meet with proper appreciation by the world, and I hope that in this account of his life I have thrown some light on this. Biographies often have an introduction in which the author protests the subject to have been "denied their proper rights by history," or declares that the "record is going to be put straight." This is that part of the introduction.

The medical historian Ralph Colp wrote, "The great English physiologist discovered hormones and the Law of the Heart. Although his name is remembered by his scientific successors, it has been forgotten by history to a curious degree" (Colp, 1951). True, though not totally forgotten, for two monographs—by Carleton Chapman (1962) and Jens Henriksen (2000)—give excellent summaries of Starling's life. In this book I have had the extra luck to access over a hundred Starling family letters, and the recollections of Ernest's grandson, Tom Patterson. I hope that these add another dimension to the story. After the beginning of the Great War there is a good deal more fine-grained detail of his life; it may be that his relatives, realizing that they had a famous man in their midst, began to keep letters relating to him. Thus the first consistent run of Starling's letters that survives was written to his mother during 1916, when at 50, he was a lieutenant-colonel in charge of gas training for the British Army in Salonika. From then on, he comes much more into focus: outspoken, and driven to fury by the medical establishment and the government.

Enquiry soon shows Starling's life to be bursting at the seams: his research, university politics, his anger, his strong views on medical education and the reorganization of UCL, his passion for mountains and his family and friends. There are two ways of dealing with such factual density—first to take each facet of his life, and deal with each ("scientist," "politician," "family man," etc.) separately from beginning to end. This is how the two previous accounts have chosen to cope. For myself, I find it disconcerting to have to switch back to the beginning of his life after each section. (Are we sure that all these Starlings are the same person?) So I have chosen a second, riskier way: to try and keep up with all the Ernest Starlings more or less simultaneously. I hope that the overall shape of his life is clearer this way, and, when we get to the end, it really *is* the end. Incidentally, the death of this remarkable man is strange, and not really explained.

Acknowledgments

I would like to thank the librarians and archivists of:

The British Library

The Wellcome Library for the History of Medicine
(especially the Contemporary Medical Archives Collection [CMAC])

The Royal Society of Medicine

The Special Collections Library, University College London

The Records Office, University College London

The Royal Society, London

The Wills Library, Guy's, Kings and St. Thomas's School of Medicine

The Public Records Office, London

The Karolinska Institute, Stockholm

The Royal College of Physicians, London

I am also indebted to Tom Patterson (Starling's grandson) and Phillida
Sneyd (his grand-niece) for family letters and photographs.

I also had all sorts of help from Andrew Baster, Paul Fraser, Jens Henriksen,
Jeffrey House (OUP), Sir Andrew Huxley, Ann-Margaret Jörnvall, Fatima
Leitao, Rodney Levick, Eileen Magnello, Inga Palmlund, Henry Purcell,
Constance Putnam, Paul Richardson, Brian Ross, Paul Sieveking, Tilly Tansey,
Andrew Thompson, Ionis Thompson, Lady Lise Wilkinson, Catherine Wilson,
and (especially) my wife, Jan, who was heroic with the manuscript.

Contents

Contents

Chronology

1915–17   Major, Captain, then Colonel in the Royal Army Medical Corps,
          involved in gas defence.

1917–18   Chairman of the Royal Society Food (War) Committee overseeing
          the country's nutrition.

1920   Operated on for cancer of the colon. Two episodes of pulmonary
          embolism following the operation

1921–26   Investigated function of the kidney—notably with E. B. Verney

1923   Appointed first Foulerton Research Professor of the Royal Society

1927   Died at sea and buried at Half Way Tree, Kingston, Jamaica

List of Illustrations

A LIFE OF
Ernest Starling

Prelude

Ernest Starling was very much a product of British nineteenth century medicine; it was a century that witnessed profound changes in both medicine and medical research. Reviewing three relevant themes—medical education, the development of University College London, and the growth of physiology—provides us with a useful backdrop to Starling's life and times. They are themes that pervade his professional career and, not surprisingly, often re-emerge in this book.

Nineteenth-Century Medical Education

Until about the middle of the nineteenth century, English medical education was in a dire state. A good deal of the blame for this can be held at the door of the two English universities, Oxford and Cambridge, which did not acknowledge science to be a suitable subject for a university education. Even if a student was determined enough to prise a medical education out of these establishments, he would have been required to provide himself with a degree in classics, and be a member of the Church of England. Thus, in 1825, when Charles Darwin wanted a medical qualification, he went to Edinburgh University. There was nothing unusual about that, for between 1801 and

1850, Oxford and Cambridge produced 273 medical graduates, while the Scottish Universities (Edinburgh, Glasgow, Aberdeen, and St. Andrews) produced 7,989 (Robb-Smith, 1966).

Things were better at the teaching hospitals of London, but only marginally. These institutions labored under the handicap of teaching by apprenticeship: students paid their chosen clinical teachers for certain courses, an arrangement that inevitably produced extraordinary inconsistencies between students. Because teachers (the consultants) ran the teaching hospitals for the benefit of their own pockets, the pre-clinical subjects—anatomy, chemistry and physiology—tended to be seriously neglected. There were no teachers specifically employed to teach what would now be called the scientific basis of medicine. When these subjects were provided at teaching hospitals, they were taught by lowly part-time clinicians, whose knowledge of basic science was often woeful. Not surprisingly, entrepreneurs saw their chance, and set up private medical schools in direct competition with the established ones, undercutting their price (Cope, 1966). Some of the teachers at these strange institutions were famously good, which added to the bizarre situation.

Thomas Huxley (who qualified in medicine from Charing Cross Hospital in 1846) began his medical education at Sydenham College, one of these private medical schools, in 1841 (Desmond, 1997). Here is the great man reflecting in later years on the awfulness of the system:

> It is now, I am sorry to say, something over forty years since I
> began my medical studies; and, at that time the state of affairs was
> extremely singular. I should think it hardly possible that it could
> have been obtained anywhere but in such a country as England,
> which cherishes a fine old crusted abuse as much as it does port
> wine. At that time there were twenty-one licensing bodies—that is
> to say, bodies whose certificate was received by the state as
> evidence that the persons who possessed that certificate were
> medical experts. How these bodies came to possess these powers
> is a very curious chapter in history, in which it would be out of
> place to enlarge. They were partly universities, partly medical
> guilds and corporations, partly the Archbishop of Canterbury . . .
> there was no central authority, there was nothing to prevent any
> one of the licensing authorities from granting a licence to any
> one upon any conditions if thought fit . . . It was possible for a
> young man to come to London and to spend two years and six
> months of his compulsory three years "walking the hospitals" in
> idleness or worse; he could then, by putting himself in the hands
> of a judicious "grinder" [i.e., a private medical school] for the
> remaining six months, pass triumphantly through the ordeal of
> one hour's *viva voce* examination, which was all that was necessary
> to enable him to be turned loose upon the public, like death on
> the pale horse . . . . (Huxley, 1893b)

The middle of the nineteenth century was, fortunately, a watershed. For during the second part of the century a number of improvements occurred—most of them anticipated by Huxley. Thus, in 1858, the Medical Act was passed. First, this established the Medical Register, a list of people recognized as medical practitioners by the state. Second, the act created a central body—the General Medical Council—among whose tasks was to ensure uniform standards of medical qualification. The act had far-reaching effects, but it did not solve a serious problem—the organization of the medical curriculum. The teaching of preclinical medicine needed to be treated as university subjects rather than master–apprentice transactions. It should have been achieved relatively easily, for successful models existed in Scotland, France, and Germany, where each subject in the medical curriculum was taught by a university department, which was usually run by a professor.

The Origins of University College

As early as 1828 these self-evident truths were appreciated by the founders of the University of London (later University College) on Gower Street (Harte and North, 1991). The original syllabus in the 1826 prospectus listed 31 proposed subjects, the last seven of which were medical: Anatomy, Physiology, Surgery, Midwifery and Diseases of Women and Children, Materia Medica and Pharmacy, Nature and Treatment of Diseases, together with Clinical Lectures, "as soon as an hospital can be connected with this establishment." Each of the seven subjects would be taught by an academic department and a nearby hospital would ensure interchange between preclinical and clinical departments. Known as the North London Hospital, it was opened in 1834, on the opposite side of Gower Street from University College London (UCL).

A simple, practical scheme, it would seem; but the resistance to it was enormous. To begin with, London University showed no affiliation to any particular religion, so it was seen as a godless competitor by Oxford and Cambridge. The London medical establishment of the time (i.e., the consultants) saw their power being eroded by autonomous pre-clinical departments. Rickman Godlee, a surgeon at University College Hospital, said "We shall lose all control over the teachers of these subjects [anatomy, physiology and chemistry] and must take care lest they keep the student too long upon the beggarly rudiments, at the expense of the essentials" (Harte and North, 1991). This book is the life of a man who spent a good deal of his life at University College, engaging the beggarly rudiments with remarkable success.

Not surprisingly, progress in the new university was depressingly slow, although the medical faculty got off to a promising start, with 54 students entered in the first year. The annual target for the college was actually 2,000 students (all subjects), but this wasn't achieved until well into the twentieth century. Money was raised by selling £100 shares (the shareholders were known as "proprietors"; a very un-university word). Other institutions, which believed in God and supported themselves with endowments and charity, were scornful of poor little University College, running on vulgar commercial money. But the idea of the place must have been attractive to teachers, because the academic standard of many of the earlier staff members was exceptionally high. Thus, Sir Charles Bell (1774–1842) from Edinburgh was a big fish for the college to catch; he was Professor of Physiology and Surgery, and gave one of the inaugural lectures. However, controversy was never far away, and in 1830–31 the Professor of Anatomy, Granville Pattison, was the cause of an extraordinary wrangle. The students complained about his incompetence, and with some reason, objected to his lecturing in hunting clothes. This led to a student demonstration, the dismissal of Pattison, the resignation of several professors (including Bell) and the resignation of Leonard Horner, the first and only warden of the college.

Physiology—the study of the normal activity of the body—was in a very poor state. In Germany, where physiological studies were far in advance of England, the approach was to explain the body's normal workings in terms of physics and chemistry (reductionism). In England there seems to have been an unspoken belief in vitalism—an assumption that science could never explain life in materialistic terms. This presumably had some sort of religious basis, but one must remember that the Victorian age was "an age of faith and of corrosive doubt; of stifling conventionality relieved by unconventional practices; of apparent conservatism yet equally of liberation and radicalism" (Ashton, 2000). The faith and corrosive doubt was epitomized by the concept of evolution and the explanation of living processes by chemistry and physics. Thus, Marshall Hall (1790–1857) an Edinburgh medical graduate who taught at Sydenham College (the private medical school that Huxley attended) proposed the spinal cord to be the anatomical basis of the reflex arc (Clarke, 1972). This provided, for example, a basis for the mechanism of sneezing, coughing and swallowing, ideas which are nowadays part of the fabric of medical knowledge. But Hall's ideas roused passionate opposition, for reflex systems excluded the soul, which was believed to be essential for many of the body's activities. Man was, in Hall's world, a machine, a mere automaton, and such ideas were unacceptable. Not surprisingly, Hall's conclusions were better received in Germany than they were in England. (Modern theology would presumably have no problems with the co-existence of the soul and the exquisite switch-gear of the spinal cord, for example.)

Central to this story is the figure of William Sharpey (1802–1880) who was appointed to the chair of General Anatomy and Physiology at University College in 1836. Sharpey's knowledge of who was doing what in physiology was extraordinary; he spent many months travelling round the laboratories of Europe (Sykes, 2000), and became a personal friend of many of the luminaries of his time. He would be seen nowadays as an arch networker. It is hardly

Figure P-1. William Sharpey: his portrait in UCL. He was Professor of Anatomy and Physiology at UCL 1836–74. (*Wellcome Library, London, with permission*)

surprising that he was Biological Foreign Secretary of the Royal Society for nineteen years (1853–72). It is interesting that Bell's and Sharpey's job titles included "physiology" as a half-subject, a situation that was shortly to change as physiology grew into a whole subject. "General Anatomy," in Sharpey's case was histology; he somehow managed to inspire several generations of outstanding physiologists, without ever doing conventional physiology. Here is one of his protegés:

> It is true that [Sharpey's] lectures were largely anatomical, that
> he carried out no physiological researches, that he performed no
> experiments on muscle and nerve other than those that had been
> performed by Galvani half a century earlier; that he never
> possessed a kymograph—the working of which he would illustrate

to his class by revolving on the lecture table what Michael Foster called "his dear old hat." (Sharpey–Schafer, 1927)

So wrote Sir Edward Sharpey-Schafer: Edward Schäfer was a pupil and Sharpey's successor at University College, and in 1918 added Sharpey's name to his own in an extraordinary and confusing act of reverence. (He also abandoned his umlaut: what Englishman would want to be thought German in 1918?) For the sake of simplicity, Sharpey-Schafer will be referred to in this book by the name he was using at the time.

It is no exaggeration to describe Sharpey as the father of English Physiology. In 1869, he and Thomas Huxley organized at University College the

Figure P-2. Edward Sharpey-Schafer (1850–1935) who changed his name from Schäfer in 1918. He was succeeded by Starling as Jodrell Professor of Physiology at UCL in 1899. (*Family collection*)

first practical biology classes in the country (Huxley, 1893a). And these two were responsible in 1876 for making physiology a compulsory part of the medical curriculum—which was a big step forward for a beggarly rudiment. Two more of Sharpey's pupils, John Burdon Sanderson and Michael Foster, became the first professors of physiology at Oxford and Cambridge respectively, thereby introducing some scientific rigor into the two courses. Some of Sharpey's physiological progeny, many of whose names will reappear in this book, are listed in the diagram below. Ernest Starling's place in the scheme is not clear-cut, though we can draw dotted lines from Edward Schäfer, whose advice he took at the beginning of his career, and William Bayliss, with whom he researched for many years. These men were the first successful (and full-time) English physiologists, and most of them were medically qualified. They were all Fellows of the Royal Society, and five of them (Sanderson, Foster, Sharpey-Schafer, Bayliss, and Lister) were knighted.

It could be said that physiology was becoming respectable at this time, were it not for the views of those sections of the public who equated physiology and medical research with "vivisection." In their eyes, physiology was far from respectable. The animals used in physiological experiments were anesthetized, a fact which antivivisectionists could not or would not accept. The British have an extraordinarily ambivalent attitude towards animal suffering: one section of the population showing indifference or pleasure toward animal suffering in the countryside (where it is often called "sport") while another section shows deep concern over the cruelty envisaged in medical laboratories. In the 1870s there was a great surge of physiological activity in England, and it had become clear that asking physiological questions was a very successful line of scientific enquiry. Sharpey's pupils were making fundamental discoveries about how the body worked, and for the next fifty years or so physiology was to become dominated by English scien-

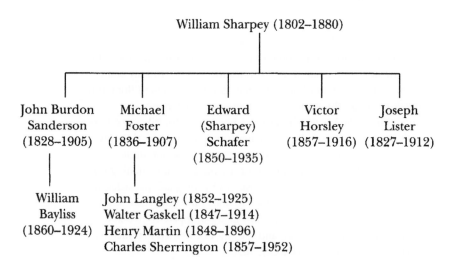

tists (not German or French ones, as it had been for the first three-quarters of the century). However, the burgeoning of physiological science in England was accompanied by a corresponding increase in the activities of Victorian anti-vivisectionists. The movement was more active in Britain than in any other country, and this is probably still true today.

The Founding of the Physiological Society

It is not to be wondered at that the general public (ignorant of the fact that no science can advance without experiment, and that for a science dealing with life, experiments must be made upon the living subject; also not knowing that in the great majority such experiments are performed painlessly under the influence of anaesthetics) was led, by the blood-curdling tales broadcasted by the anti-vivisection agitation, to voice a demand for the regulation or even the abolition of experiments on living animals. (Sharpey-Schafer, 1927)

However quirky Sharpey-Schafer's prose, there is no doubt of the seriousness of his message. For in 1875, a Royal Commission on Vivisection, presided over by Lord Cardwell, was established. Thomas Huxley and John Erichsen (Professor of Surgery at University College) sat on the Commission, representing Physiology (Harte and North, 1991). The Commission recommended that vivisection should be regulated by an act of parliament, and that only scientists licensed by the Home Secretary should be allowed to do such experiments. Furthermore if experiments involved operating on unanesthetized animals, special certificates would be required.

Huxley's thoughts on this were characteristic:

But while I should object to any experimentation which can justly be called painful, for the purpose of elementary instruction; and, while, as a member of the late Royal Commission, I gladly did my best to prevent the infliction of needless pain, for any purpose; I think it is my duty to take this opportunity of expressing my regret at a condition of the law which permits a boy to troll for pike, or set lines with live frog bait, for idle amusement; and, at the same time, lays the teacher of that boy open to the penalty of fine and imprisonment, if he uses the same animal for the purpose of exhibiting one of the most beautiful and instructive of physiological spectacles, the circulation of the web of the foot. No one could undertake to affirm that a frog is not inconvenienced by being wrapped up in a wet rag, and having his toes tied out; and it cannot be denied that inconvenience is a sort of pain. (Huxley, 1893a)

Physiologists felt threatened by the act, though at this distance in time, the proposals seem far from draconian. On March 28, 1876, Burdon Sanderson wrote to Schäfer (Sanderson, 1867).

My dear Schäfer

It is proposed to hold a meeting at my house at 5.30 p.m. on Friday next of a preliminary character for the purpose of considering whether any, or what, steps ought to be taken with reference to the recommendations of Lord Cardwell's Commission. It will probably also be proposed at the meeting to form an Association of Physiologists for mutual benefit and protection. Sharpey, Huxley, Foster, Lewes and others have promised to attend. I shall be glad if you can come also.

Yours very truly, J. B. Sanderson

(Being the days before duplicating, the invitations were all written out in longhand by Sanderson's wife.)

So the meeting went ahead at Queen Anne Street, London, *chez* Sanderson. Apart from the four mentioned in his letter, Schäfer, Francis Galton, David Ferrier, and Francis Darwin (Charles' son) were a few of the original nineteen attendees. A committee was appointed to draft a constitution, and the Physiological Society came into being. It has played a vital role in the development of the subject ever since. Oddly, the minutes of the occasion contain no reference to the Cardwell Commission; perhaps the founders were so carried away by becoming a society that they forgot the rather depressing reasons for their coming together. The minutes were signed by G. H. Lewes (the Victorian polymath who subsequently shocked society by his non-marriage to George Eliot) (Ashton, 2000). A professional society for physiologists would probably have come about sooner or later, but the anti-vivisection movement certainly provided a trigger for the society's birth in 1876.

The original meetings of the Society began with dinner at 6 o'clock in the evening, for it was a dining club (Minute Books of the Physiological Society). Scientific communications and demonstrations, which today make up the bulk of a meeting, were after-thoughts to the original design of meetings. The first constitution of the society proposed no more than 40 members, a rule that changed very quickly, for over a century later, there are around 2,500. It is ironic that such a thriving organization was born out of the original members' fear for their livelihood. "Ex malo bonum" (good out of evil) as Sharpey-Schafer comments in his chatty history of the society (Sharpey-Schafer, 1927). The society's membership first included women in 1915 (Tansey, 1993).

Many years later, Sir Jack Eccles analyzed the success of the society:

As I survey in retrospect this amazing development of British physiology during the one hundred years from 1870, I am

convinced that the Physiological Society has been the greatest
factor in this success. In fact I regard it as an ideal of what a
scientific society should be. It is informal, yet it is efficiently
organized. There is an excellent atmosphere of personal friendli-
ness, yet those presenting scientific papers or demonstrations are
subject to the most searching criticisms; what surprises visitors is
the tradition that these criticisms must be free of personal attack
and that they must be accepted with good grace. There is a
characteristic sporting atmosphere in the way the hard knocks are
given and received. (Eccles, 1971)

There was no British publication given over solely to the subject, but as it
gathered momentum, British physiology needed its own mouthpiece. In
1885, the *Journal of Physiology* was founded by Michael Foster in Cambridge.
This title had a certain intellectual arrogance, for French and German
journals had existed for some time, and their titles included their country
of origin. (This is reminiscent of British postage stamps that also find it un-
necessary to state their nationality. Could this also be Victorian arrogance?)
When a rival journal was established across the Atlantic in 1887, it was the
"*American*" *Journal of Physiology.*

Over the succeeding century the *Journal of Physiology* has gone from strength
to strength, although its official title is now the *Journal of Physiology (London).*
Many of the topics discussed in this book first appeared in its pages. We will
continue our description of the growth of the subject after we have introduced
Ernest Starling.

1

Hearts and Capillaries

Ernest Starling's Arrival

During Queen Victoria's reign, the British Raj provided employment for a marvellously wide range of expatriates. Among them was a barrister, Matthew Henry Starling, who was called to the bar in London in 1863, went to India in 1868, and spent the rest of his legal career in Bombay. There he served first as an advocate, and in 1887 was made clerk to the crown court for the rest of his life. On two occasions he was acting ("puisne") high court judge (in 1895 and 1902) and wrote a standard textbook, *Indian Criminal Law and Procedure*, in 1869, which ran to five editions. It seems curious that a man with such a grasp of Indian law should only have been a locum, a stand-by, judge. He also seemed to be a stand-by husband; for in 1864 he married Ellen Matilda Watkins, of Islington, and left for India four years later, leaving her in Islington. His post allowed him a journey home every three years, but his absence didn't stop this enterprising couple from producing seven children over eleven years; six of the children were born in Islington and one in Bombay. How often did Ellen sail to India? We are unlikely to discover, but Theodora (born in 1865), Ernest Henry (1866), Gertrude (1867), Bernard (1869), Mabel (1872), Hubert (1874), and Bertha (1876) bore witness to the success of their parents' social arrangements.

We have a striking photograph of Matthew Starling, where, looking like a villain in a silent film, he is perched on a wonderful machine that is a penny-farthing tricycle. He died in 1903, at age 65 (of "cardiac syncope" according to his death certificate) in Bombay. Presumably it took several weeks for the Islington team to hear the news.

By all accounts, Ellen Starling was a remarkable mother. She brought up her middle-class family on strict Anglican principles, and her eldest son Ernest seemed particularly close to her (they wrote affectionate letters to each other for the remainder of Ellen's life). Ernest was born at 2, Barnsbury Square, Islington, on April 17, 1866, and, as the family expanded, it moved to a larger house (32 Milner Square) nearby. In 1881 the household con-

Figure 1-1. Matthew Henry Starling, Ernest's father, in Bombay in the 1890s. (*Wellcome Library, London, with permission*)

Figure 1-2. Ellen Starling and her children, taken about 1877. From top left, clockwise, the children are: Theodora (born 1865), Ernest (1866), Mabel (1872), Hubert (1874), Gertrude (1867), and Bernard (1869). Bertha (1876) is on her mother's lap. (*Family Collection*)

sisted of Ellen, her seven children, Ellen's unmarried sister (Jessie Watkins), and a maid (Jane Fowler).

As the household's alpha male, Ernest had unusual social responsibilities and he seemed to have coped well. From 1872 to 1879 he went to the Islington Proprietary [sic] School (fee-paying) and in January 1880, when he was thirteen, to King's College School, which was then in the Strand. The subjects he studied included divinity, Greek, Latin, French, ancient history, English, mathematics, and chemistry. He won the Middle and Upper Fifth Literature Prize, the special Greek Prize, and the Council's Chemistry Prize. Sadly, these rather bare facts are all we have about his youth and education. It is not clear exactly when Ernest decided that he wanted to study medicine; but he had an uncle who had qualified at Guy's Hospital, and in 1882, at age 16, he became a medical student there. Ernest's younger brother Hubert entered Guy's Hospital as a student eight years later.

Guy's Hospital

The conventional word for academic careers like Starling's is "glittering." His contemporary and friend at Guy's, Charles Martin, estimated that Starling won two-thirds of the prizes that were available, and that if he had melted down the gold medals awarded to him he would have "enjoyed relative affluence." He doesn't seem to have melted them, however, for the next few years were spent in relative poverty. In his early days at Guy's his ambition was to be a Harley Street physician, but within a year or two he began to doubt this rather conventional ideal, and wondered whether he might devote himself to an academic career. "For as soon as he touched the study of natural science, it was clear that Starling had found his métier. The causal relation of facts enthralled him" wrote C. J. Martin in 1927. He was bowled over by the sweep of biological ideas—evolution, obviously, but also by the relatively new world that was explaining life in terms of chemistry and physics.

Starling actually wrote an article in 1887 for the hospital magazine, *Guy's Hospital Gazette*, on heredity. The article is scientifically respectable, defending Weismann's germ-plasm theory and attacking Lamarckism. But he received a strong letter in reply from J. R. Ryle, a Guy's graduate who was in general practice. Ryle was a Lamarckist, and made use of gout to demonstrate Lamarckian ideas ("It is well-known that over-indulgence at the table will set up gout in a man's descendants"). Clearly enjoying a fight, Starling replies:

> Since reading Dr Ryle's letter I have looked up all the authorities
> on gout that I could find in our library, and must say again that I
> can find no evidence that too much indulgence in the pleasures
> of the table will set up gout in the man's descendants . . . if Dr.
> Ryle, instead of mentioning 'hundreds of well-observed cases' in a
> vague sort of way, would give us the records of half-a-dozen
> typical cases which would prove his case conclusively, I should be
> the first to give up Weismann's theory in its present condition.

Undaunted, Ryle replies; but he is only muttering to himself, and Starling sees no need to continue. The vigor with which the student Starling defended his views is absolutely characteristic; speaking his mind was already a strength and a weakness in his character.

An important influence at this time was his friend Leonard Wooldridge (1857–1889), a young physiologically oriented Guy's physician. Wooldridge had a star-studded academic career through Guy's, and in 1879 went to Leipzig, where he worked for a year with Carl Ludwig, the leading Ger-

Figure 1-3. Ernest Starling, photographed in Germany, aged about twenty. (*Family collection*)

man physiologist of the time. There, Wooldridge became engaged to Florence Sieveking, an English visitor who was studying German and music, and married her in 1884; this is of great relevance, as we shall soon see. Wooldridge persuaded Starling (now half-way through his medical course) that he, too, should visit Germany, and Ernest spent the long vacation of 1884 in Willy Kühne's laboratory at Heidelberg, where he became more firmly convinced that he was going to become a physiologist. He returned, bilingual and teutonized, with his hair "en brosse" (like a brush), to finish his clinical medicine career. In his medical finals in 1888 he won the University gold medal for medicine, which probably came as no surprise to anyone. Then there were two house jobs at Guy's—one with William Hale-White, a young consultant physician who became a lifelong friend of Starling's, and who later remarked that "he had a marvellous flair, not only for the science, but also for the art of clinical medicine." While working for Hale-White in 1889, Starling caught diphtheria from a patient, and for a while his survival was in doubt. But there is no suggestion of the disease leaving any permanent damage, for throughout most of his life he seems to have been a person of extraordinary physical energy.

There was however the difficult matter of becoming a physiologist. At Guy's there *were* no professional physiologists—Wooldridge was a consultant physician with a passion for the subject and there were two other part-timers. H. C. Golding-Bird (a surgeon) and J. W. Washbourn (a bacteriologist and physician) lectured to students in the two preclinical years. Golding-Bird's major contribution to the subject was his skill at cutting microscopic sections with a razor, perhaps not quite such a strange talent as it appears, for the study of the structure of tissues (histology) made up a significant part of the curriculum.

In 1887 Golding-Bird proposed to the Medical Committee of Guy's Hospital the building of two laboratories devoted solely to physiology. The proposal was seconded by Wooldridge, and was passed. In 1888, *Guy's Hospital Gazette* included an editorial:

> At last Guy's is to have a Physiological laboratory. Though late, we
> are not last. There are still some hospitals in London where such
> an institution does not exist. A wooden fence at present marks
> out the site of the future laboratory, a good north light is
> available, of which we trust, full use will be made, as it is impos-
> sible to do good microscopical work with a top-light alone, as well
> seen in the present class-rooms. [Electric lights were not available
> for another fifteen or twenty years.]

The laboratories were finished by September 1888. They were actually built on a site behind what is now the Nuffield Private Patients' wing.

Against this background of improving facilities Starling became a Demonstrator in Physiology in 1889. He was the only full-time lecturer; he must have been very tempted to be a half-and-half teacher like his peers, but his determination and belief in himself were remarkable. He went for help to Professor Edward Schäfer at University College. This is the earliest Starling letter that has survived (Starling, 1891).

> Dear Professor Schäfer   May 21, 1891
>   I am glad to say that I was partially successful in the matter of
> lectureships.
>   The Guy's staff could not make up their minds between
> Washbourn and myself, so they appointed us both. As Golding-
> Bird only gives up to one fourth of the lectures, Washbourn and
> I each take one eighth. It is a rather ridiculous arrangement,
> but I suppose the title of Joint Lecturer will be useful to me—at
> least I hope it will. It certainly is not much of a catch from a
> pecuniary point of view.
>   I am so deeply in your debt already that I feel quite ashamed of
> asking you to help again.

I wrote last night applying for the British Medical [Association] research scholarship which is advertised as vacant.

I want to come and see you and talk over the subject I have sent in. It is so difficult to put down ones ideas for future work on paper, as they are constantly changing.

With many thanks

I am

Yours very sincerely

Ernest H. Starling

Applying to the British Medical Association (an unlikely source of funding nowadays) he was given a scholarship, which was £150 a year and kept him above the breadline. At this time he had published no scientific papers.

In May of a remarkable year Wooldridge ate a late lunch in Guy's that, he believed, contained some elderly fish. He had diarrhea and vomiting for a few hours, was then unwell and, in spite of treating himself with a day's train trip to Hastings, died a week later. A post-mortem showed extensive ulceration of his large and small bowel, "enteritis" being given as the cause of death (could it have been typhoid?). His death, at 31, was a great shock to Guy's, as his obituary in *Guy's Hospital Reports* concluded: "His life was as happy as it was brief. He used the talents entrusted to him with rare fidelity, and his brief and brilliant career shows once more that much may be accomplished in few years: 'And in short measures life may perfect be,'" (Pye-Smith, 1889). *Guy's Hospital Reports*, now sadly defunct, had earlier published the original observations of Addison, Bright, Hodgkin, and other Guy's celebrities.

Wooldridge had published twenty-six papers (sixteen in English, ten in German); most of them involving the coagulation of blood. A collection was made for a memorial (Starling contributed £5, a huge sum considering his income). The resulting art nouveau plaque, featuring a mournful nymph, can be seen today on the outside wall of the Harris theatre on the Guy's campus.

Starling took over Wooldridge's teaching tasks in physiology, but somehow this failed to improve his finances, for Washbourn and Starling each received one half of a putative lecturer's salary. Starling lived in the hospital, which suggests a certain desperation. Guy's support of the developing subject of physiology was woeful—the clinicians were displaying the "beggarly rudiments" syndrome that we saw earlier.

Much later in his life, Starling recalls his first days as a physiologist. This is part of a letter to Charles Lovatt Evans:

When I was appointed at Guy's . . . the only physiological laboratory was a small empty room. When I applied for £200 to buy apparatus for an advanced class, as well as for my private work, I was informed that "a medical school was not a place to do research in." The only thing that mattered was to get the students through their examinations. When I arranged my work so as to have one summer free to go abroad and learn something of the subject I was teaching, I was warned that I must not expect reappointment if I took such an unwarranted step. . . . As a matter of fact they promoted my junior to my place. (Starling, 1919)

The year 1890 saw the publication of Starling's first scientific paper, "A note on the urine in a case of phosphorus poisoning" published in *Guy's Hospital Reports* (Starling and Hopkins, 1890). Starling's paper, hardly a serious contribution, is memorable for two reasons. First, the patient died after eating "a slice of bread and butter which had been thickly coated with phosphorus paste." (It was rat poison, presumably.) Second, Starling's coauthor was Frederick Gowland Hopkins, who became the most distinguished English biochemist of his time, and won a Nobel Prize in 1929. It was Hopkins's and Starling's only joint publication, though they often did experiments together in the 1890s.

The ambitious Starling was soon aware of Guy's shortcomings as a centre of physiology. Not surprisingly, he began to visit Schäfer's department at University College, where he undoubtedly found a better class of physiological conversation. There in 1890, he met the affable man who was to become his friend, brother-in-law, and coresearcher for the next twenty-five years: William Bayliss.

William Bayliss

William Maddock Bayliss came from an unlikely background. He was born in 1860 in England's industrial midlands, in Wednesbury, between Birmingham and Wolverhampton. His father, Moses, was an up-market blacksmith, and helped found the successful firm of Bayliss, Jones and Bayliss, "Makers of Bolts, Nuts, Screws, and Ornamental Iron Gates." William was in effect an only child, for although his father had several children by four wives, William was the only one who reached adulthood. He was educated at a private school in Wolverhampton, and from an early age showed great skill at making things with his hands. He did this for a short while with Bayliss, Jones, and Bayliss, but it wasn't a success. He then became apprenticed to a local medical practitioner, with whom he studied dressings and dispensing.

But, in 1880, Moses retired from the business and did what seems an extraordinary thing. He took his wife and son to the outskirts of London, where he built a large house between Golders Green and Hampstead, then a charming rural area. The house, St. Cuthberts, in West Heath Road, had about four acres of garden, and was next door to an even bigger garden belonging to Sir Thomas Spencer Wells (1818–1897), surgeon to Queen

Victoria's household, whose name will always be associated with the artery forceps that he invented.

In 1880, William attended University College to enter the medical course. William passed the preliminary medical exam—chemistry, physics, botany and zoology—in 1881, and was awarded an Entrance Exhibition. But in doing the medical course, he found that anatomy, with its mind-numbing detail, was too much for him; he failed the exam and had to drop out of the course.

But he was greatly attracted to physiology, and was taken with the teaching of John Burdon Sanderson, who was at the time professor of physiology at University College (he had succeeded William Sharpey in 1874). In 1883, Burdon Sanderson became the first professor of Physiology at Oxford, and as a student Bayliss followed him two years later, reading physiology at

Figure 1-4. The Bayliss home, St. Cuthbert's, on the edge of Hampstead Heath, taken in the 1890s. Note the horse and carriage. The figure in the foreground seems to be raising his hat to the photographer. The house is no longer a private residence, but is now the headquarters of a religious organization, and includes some unfortunate additions to its gothic Victorian architecture. (*Wellcome Library, London, with permission*)

Figure 1-5. William Bayliss, photographed in 1878, aged 18. He was six years older than Starling. (*Wellcome Library, London, with permission*)

Wadham College. There he was rather older than his contemporaries (he was 25) and known as "Father Bayliss." His son, Leonard, writes that his fatherly air was heightened by a bushy beard, for throughout his life William never shaved. He obtained a first class degree and returned to University College London in 1888, becoming "assistant" in the physiology department, under Schäfer. He stayed at UCL for the rest of his life.

There was virtually no accommodation at the college at the time, and William lived at home. Travelling to work involved a long walk across the heath to Hampstead, and then a horse-bus to within a few minutes' walk of the college. (The journey became simpler in 1905, when the Northern Line was built, for St. Cuthberts was only ten minutes' walk from Golders Green Underground Station.)

Thanks to Bayliss's son Leonard, we know rather more about Bayliss's early life than Starling's (L. Bayliss, 1960). When the two met in 1890, Bayliss had three or four years start as a practicing scientist. It is easy to imagine their meeting, for Bayliss was the kindest and most hospitable of men, and would have welcomed the rather gauche enthusiast from Guy's. Bayliss was cautious, and well-read; he enjoyed the technical side of experiments. Starling was impulsive, less knowledgeable at this time, and only interested in experimental techniques in so far as they gave answers to the immediate problem. (Bayliss was also very hairy and Starling clean-shaven, and while we are playing this game of opposites, Bayliss was very well-off and Starling impoverished.) It would have been no surprise if they had shown an instant antipathy for each other. But it was not so: they talked and argued endlessly and passionately, but with no animosity. Everyone writing about their relationship commented on their disparate (but complementary) personalities. Thus Chapman wrote "that the one needed the other there can be little doubt, but the astonishing thing, in this day of claim and counterclaim for credit and priority, is that serious friction between the two seems to have been minimal" (Chapman, 1962).

Both men had been guests at meetings of the Physiological Society in the late 1880's, and on February 15, 1890, at a meeting in University College, they were simultaneously elected members. This is recorded in the minute book in the elegant handwriting of Charles Sherrington, who had recently been elected secretary. Bayliss became an indefatigable attender of the Society's meetings—Starling less so. Some years later, when Bayliss was invited to Buckingham Palace to receive a knighthood, his first reply was that he couldn't go to the ceremony because there was a meeting of the Physiological Society that day.

Early Research—Recording from the Heart

As Bayliss was the more experienced scientist, it is hardly surprising that the pair's first experiments reflected his interests. Having come under Burdon Sanderson's influence at Oxford (whose interests were electrophysiology and the heart) their first foray was almost inevitable—the recording of electrical activity from the heart.

The heart's action consists of the auricles (atria) contracting (and pushing their contents into the ventricles) followed almost immediately by the ventricles contracting and pushing their contents into the lungs and the rest of the body. For proper functioning of the heart the two sets of contractions have to be in the right order and have the proper timing. The contractions are driven by an electrical wave that originates at the back of the auricles, passes down between the ventricles and then spreads over their surfaces.

The wave has a function analogous to the current passing to the spark plugs in an internal combustion engine. Without proper timing of this ignition system, the heart cannot function. The wave can be electrically recorded from the beating heart itself, or, using a more sensitive detection device, from the surface of the body related to the heart.

The electrical detection device that Bayliss and Starling used was the capillary electrometer, an instrument invented by Gabriel Lippmann (a Frenchman working in Gustav Kirchhoff's laboratory) earlier in the century (Frank, 1988). Lippmann put mercury in a vertical capillary tube, and covered the surface (meniscus) of the mercury with a conducting solution (he used dilute sulphuric acid). When an electric potential is applied across the mercury and the acid, the shape of the mercury meniscus instantly changes. In a fine capillary tube, the tip of the mercury actually moves upward. A beam of light passing through the capillary will be interrupted by the mercury, and if the beam is focused on a moving photographic plate, a record of the rhythmically changing electrical activity results. The plate was made to move by a Heath Robinson railway system that had been invented by Burdon Sanderson. As can be imagined, the capillary electrometer was a very cumbersome creature. University College had one, and its light source was theatrical limelight (which was not very bright, and limited the sensitivity of the instrument). At Oxford, Burdon Sanderson also had one; his was more sensitive, having a brighter light source—an oxy-hydrogen arc.

Bayliss and Starling investigated the electrical activity of the mammalian heart (Bayliss and Starling, 1892a). Previous work had mostly used the hearts of frogs, which continue to beat outside the body when provided with a nutrient solution. Mammalian (i.e., warm-blooded) hearts stop beating almost immediately when removed from the body, and so were virtually impossible to investigate in this way. Using the capillary electrometer at University College, Bayliss and Starling looked for the electrical activity associated with contraction in the heart of the anesthetized dog. They concluded that (as expected) the electrical wave began in the auricles and spread to the ventricles after a delay, which they showed to be between 0.12 and 0.16 seconds. So a tenth of a second is all that is necessary for the atria to eject its blood into the ventricles. In the descriptions of their experiments, they speculate on what the anatomical basis for the delay might be. There was not a nerve involved, for if the electrical wave was passing from auricles to ventricles via a nerve, conduction would be much faster, and the delay would be less. Under the circumstances, the heart could not work as a pump, for the atria would not have the crucial tenth of a second in which to empty themselves into the ventricles. In fact the anatomical question had to wait some fifteen years for an answer: Sunao Tawara in 1906 demonstrated a structure—the atrio-ventricular node—containing a network of very small modified heart-muscle cells (Tawara, 1906). These tiny fibers conducted the electrical wave relatively slowly, and produced the crucial tenth-of-a-second delay.

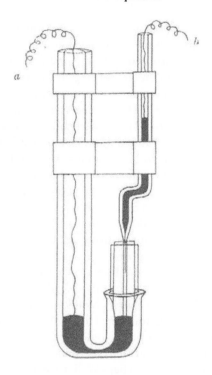

Figure 1-6. A capillary electrometer. A change of electrical potential between *a* and *b* causes the mercury meniscus in the capillary *c* to move, and this can be recorded optically. Starling and Bayliss recorded their own electrocardiograms in 1891–92, using Burdon Sanderson's electrometer in Oxford. (*Schäfer's Text Book of Physiology, 1898*)

Wanting to extend their experiments to humans, Bayliss and Starling knew that they could only record the heart potentials through the skin, which reduced the size of the potential. Needing a more powerful electrometer, they went to Oxford to use the Burdon Sanderson machine. They recorded each other's electrocardiogram—for that is what it was. The best traces were obtained by having one recording electrode over the heartbeat on the chest, and the other attached to the right hand. All these findings corroborated what they had previously found recording directly from the heart of the anesthetized dog. The delay between atria and ventricles was virtually identical to that in a dog's heart.

It would be gratifying to report that these findings were an important contribution to human knowledge. Unfortunately they weren't. Most of their results repeated the findings of the exotically named Augustus Désiré Waller (1856–1922). Waller had graduated in medicine from Aberdeen and was persuaded to go into research by Sharpey. After research in Germany, he came back to work with Burdon Sanderson at University College, and then became a lecturer at St. Mary's Hospital. He achieved an unusual degree of financial independence for his research by marrying Alice Palmer, a biscuit heiress. (Huntley & Palmer is a long-established English brand of biscuits.)

Using the capillary electrometer, Waller had provided first a verbal description of the first human electrocardiogram in the proceedings of the Physiological Society at Cambridge in 1886, and published the tracing in the *Journal of Physiology* in 1887. He emphasized that the voltages recorded from the heart were the *cause* of the contractions, and were not produced by them. On his traces the voltages always came just before the record of the heartbeat. Waller seems to have had a flair for self-advertisement, for he spent several years (on and off) showing-off his findings around Europe. It was noted that he always seemed to be showing his original (1886–87) recordings; perhaps Bayliss and Starling hoped that they could explore further than Waller. But they couldn't—their electrometer tracings look just like his, and were published four or five years later.

At this time neither Waller, Bayliss, nor Starling could envisage the enormous clinical importance that the human electrocardiogram was to have. The tracings were seen only as interesting physiological curiosities. Eighteen more years would pass until the Dutch physiologist Willem Einthoven (1860–1927) and the British investigator (Sir) Thomas Lewis (1881–1945) changed the electrocardiogram from a physiological curiosity into an essential cardiological tool. Einthoven's interest in the subject began the day that he saw Waller giving a lecture on it, so perhaps Waller's self-advertising was not such a bad thing.

Bayliss and Starling had not actually failed in their venture: they produced, in retrospect, the "right" answers. It was just that they were second, and to be second in science is to be last.

Stimulating the Heart

The second series of experiments upon which the pair embarked was concerned with the effects of nerve stimulation on the action of the heart (Bayliss and Starling, 1892b). It was known that two sets of nerves—the vagus and the sympathetic—had approximately opposite effects: stimulating the vagus slowed the heart down, whereas stimulating the sympathetic nerves speeded up the heart and made ventricular contractions more forceful.

Previous research had been done on cold-blooded animals: Bayliss and Starling were establishing that these principles applied to mammals as well (they used the anesthetized dog). Stimulation of the vagus slowed down the frequency and force of contraction of the atria; the effect on the ventricles was less clear-cut. But they showed that strong vagal stimulation may stop ventricular contraction entirely. (This is an observation of great clinical relevance, for during fainting the heart is slowed by the vagus to such an extent that blood flow to the brain is seriously hampered, giving rise to the faint). Perhaps the most original part of this paper was to show that the vagal stimulation slowed, and sympathetic stimulation speeded up, the conduction of the cardiac potential across the atrio-ventricular node. So sympathetic stimulation shortened the conduction time from 0.16 to 0.13 seconds. Such speeding up would occur, for example, in exercise, when the ventricles would fill faster than normal. The measurements were made from simultaneous recordings of atrial and ventricular contractions on a smoked drum recorder (a kymograph) without the use of the capillary electrometer. (This was because the research was done at Guy's, and Guy's did not possess an electrometer.)

Unfortunately, Bayliss and Starling were second again. In the introduction to their paper in the *Journal of Physiology* we read:

> After we had written the rough draft of this paper, an abstract of
> a communication by Roy and Adami to the Royal Society ap-
> peared. . . . As will be seen, our work confirms most of their
> results, so far as we treat of the same subjects.

Their research was competent and resourceful, but they were not pushing back any frontiers. However, facilities at Guy's were steadily improving, and this was largely a result of Ernest's enthusiasm. Hale-White, Starling's physician friend, observed: "Anyone strolling into the laboratory would find Starling, Hopkins, Bayliss, and several others all hard at work. The department became a living thing of great vitality" (Hale-White, 1927).

During 1890, Starling began writing a textbook. Few people involved in teaching a subject for less than two years would have even contemplated doing such a thing. *The Elements of Human Physiology* (1892) was, in spite of its title, no student crammer, for it was 464 pages long; nor was it an amateurish first shot at a textbook, for it ran to eight editions. The last edition was published in 1907, by which time its length had crept up to 716 pages, and perhaps outgrown its title.

Marriage to Florence Wooldridge

Socially, Ernest was keeping-up with Wooldridge's widow, Florence. In fact, he was more than keeping-up with her, because on December 21, 1891, they were married at St. Thomas' Church, Portman Square. It all seems rather abrupt, but there are no surviving family documents from this time, so we know little of the circumstances of the wedding. The marriage certificate was witnessed by Ernest's mother, Ellen Starling (Father Starling being in Bombay) and by Florence's uncle, Gustav Sieveking.

Marriage to an old friend's widow could involve a few psychological hangups; if this were so, there were no outward signs of them. Florence Amelia Wooldridge had been born in 1861, and was five years older than Ernest. She was the daughter of Sir Edward Sieveking, a physician at St. Mary's and

Figure 1-7. Florence Starling, photographed in Leipzig, where she went to have piano lessons in the late 1880s. She may have been married to Starling's mentor Leonard Wooldridge when this picture was taken. Wooldridge died in 1889, and Florence married Ernest in 1891. (*Family collection*)

one of the large number of doctors employed to preserve Queen Victoria's health. The German thread is very strong, for the Sieveking family was originally from that country, and Florence first met Leonard Wooldridge in Leipzig, where she was learning German and studying the piano. At various times in her marriage to Ernest she was to be wife and mother, secretary, translator, editor, indexer, proofreader, and accompanist; Ernest was an enthusiastic baritone, and, almost inevitably, they had a shared passion for lieder.

After the wedding they moved to 14, Grosvenor Road, Westminster, which was probably the most prestigious address that Ernest had had for a long time. They spent the summer of 1892 in Breslau (now Wroclaw in Poland), where Ernest was working with the professor of physiology, Rudolf Heidenhain. Their subject was the formation of lymph, a topic in which Ernest's interest had probably been stimulated by Wooldridge. Throughout his life Starling saw physiology as leading medicine forward, and in this early work on lymph he was seeking understanding of the mechanisms of fluid accumulation in tissues (edema; dropsy).

Offers from Oxford

About a week after Ernest and Florence arrived in Breslau, a letter came from Burdon Sanderson in Oxford. The letter has not survived, but we have Ernest's answer; Sanderson was offering him a job (Starling, 1892).

> Vorwerks Strasse 11, Breslau 20.6.1892
> Dear Professor Sanderson
> If we had got your letter a month ago, we should have jumped
> at the idea of going to Oxford—my wife and I would like nothing
> better than to stay in Oxford—and I would rather work there
> than anywhere else . . . I think it is so important not to have too
> much teaching work while I am mainly a student—that I don't
> think it would be wise now to give up my teaching post at Guy's at
> present . . .
> Our first thought was to accept your kind offer at once and I
> was very disappointed that on second thoughts it seemed wiser to
> retain my present job. Another objection to changing was that I
> had only just re-applied for re-appointment and it might give rise
> to a good deal of irritation at Guy's if I resigned the post directly
> after I had applied and been re-appointed.
> With many thanks for so kindly thinking of me.
> I am, yours very truly, Ernest H Starling.

Was Ernest not showing extraordinary loyalty to Guy's? Most people, when offered a better job, are hardly concerned with the feelings of their current employer.

Sanderson then sets his sights lower, and his next letter (not surviving) offers Ernest some lecturing at Oxford. Starling replies:

> Breslau 29/6/92
> Dear Professor Sanderson
> In the first place please accept my thanks for the honour you
> do me in asking me to take on your lectures next winter. I shall
> be delighted to do so if I can possibly arrange it . . . I have,
> however, written to Golding-Bird, who, as senior, has the choice

of times, to ask him to let me lecture for the first 6 weeks, from
the first of October. If he consents, there will be no difficulty at
all. I will let you know as soon as I hear from him . . . With kind
regards, in which my wife joins
　　Ernest H. Starling

Sanderson had presumably heard Starling speak at the Physiological
Society. There was little love lost between Ernest and his colecturers at
Guy's—it is difficult not to believe that they resented him. So his final letter
comes as no surprise:

　　　　　　　　　　　　　　　　　　　　　　　　　Breslau July 1892
Dear Professor Sanderson
　　I have only just heard from Golding-Bird. He declines to alter
the date of his lectures—"is not prepared to take up the other
branches which he had dropped"—so my six weeks lectures must
come after Christmas . . . I am very sorry. It is unfortunate that my
little six weeks at Guy's should coincide with the period you
mention at Oxford.
　　Believe me, yours very truly
　　Ernest H. Starling

Had he any inkling of the later problems he was to have at Guy's (as
anticipated here by Golding-Bird's bloody-mindedness) he might have accepted Sanderson's first offer and taken a job at Oxford, where his life would
have gone off in a quite different direction.

Lymph (1)

The main function of the circulation of the blood is to transport molecules
to and from tissues of the body, and these molecules exchange at the small-
est vessels in the circulation: the capillaries. Capillaries are easily permeated
by gases and small molecules, but the larger the molecule, the greater the
difficulty in crossing the capillary wall. This is particularly relevant for the
proteins present in blood—the plasma proteins. Thus, the fluid that passes
out of capillaries is blood without cells and most of its proteins. The fluid
passes out of the capillary because of hydrostatic pressure within the capillary (20–30 millimeters of mercury) that is derived from the contraction of
the heart.
　　If the capillary pressure is raised, the outward movement of fluid increases
(it would have made many of Starling's experiments simpler had he been
able to measure this pressure directly, but this was not achieved until about
30 years later by Eugene Landis, in Pennsylvania). Excess fluid from the
circulatory system capillaries finds its way to the capillaries of the lymphatic

system. When the fluid is in the lymphatics, it is known as lymph, but chemically it has not changed. Lymph capillaries have large pores in them and a series of valves, so that lymph can only move toward the center of the body. Lymph finally is collected into a single vessel (the thoracic duct) that passes into a large vein in the neck. A tube tied into the thoracic duct of an anesthetized animal enables all the lymph passing from the body to be collected and analyzed in an experimental period.

The problem that occupied Starling was the actual mechanism of lymph formation. There are two indisputable factors that influence the rate of lymph flow—the pressure inside, and the permeability of, the (blood) capillaries. The great German physiologist Carl Ludwig (1816–1895) had proposed that these factors were all that was needed to explain any experimental finding with lymph; this was known as the "'filtration' hypothesis." Another German physiologist, Rudolph Heidenhain (1834–1897) had performed a series of experiments whose results, he believed, could *not* be explained by this mechanism. Thus, under certain experimental circumstances (which will be explained later), he found that some substances reached a higher concentration in the thoracic duct lymph than their simultaneous concentration in plasma. For this to occur, the capillary walls would have to concentrate the relevant substances. In order to achieve this, the capillary walls would have to do work, as gland cells do when they secrete. Heidenhain's concept of lymph formation was thus known as the "'secretory' hypothesis"; it was actually fallacious, but it was an attractive fallacy, and Starling spent four or five years disproving it. In the process he actually made several major discoveries. Heidenhain had published his paper in 1891 (Heidenhain, 1891), and it is likely that this played a role in attracting Ernest to Breslau in 1892.

Starling repeated some of Heidenhain's experiments in Breslau, and obtained the same results as the German. When he returned to Guy's in the autumn, he repeated some of the others and again confirmed the results. But instinct told him that the secretory interpretation was false; he had to prove it so. Some good historical reviews of the whole topic have been published, such as H. Barcroft (1976) and Michel (1977).

Heidenhain's experiments included the effects of obstructing (occluding) large blood vessels on thoracic duct lymph flow. The vessels that he occluded were the aorta, inferior vena cava, and the portal vein. Obstructing the aorta produced an immediate drop in arterial pressure, as one would expect, but had little effect on lymph flow. Obstructing the inferior vena cava produced a fall in arterial blood pressure, and an enormous increase in lymph flow (which was richer in proteins than normal—how could *that* be explained by the filtration theory?). Obstructing the portal vein produced a small fall in arterial blood pressure and an increase in lymph flow.

Ernest spent months trying to understand why occluding the inferior vena cava should have such an extraordinary effect on thoracic duct lymph. In a resting animal, the lymph in the thoracic duct is only derived from the liver and the intestine. Heidenhain believed that it was predominantly from the intestine. But Starling noticed that when the inferior vena cava was occluded, lymphatics draining the liver were suddenly very conspicuous (they were on their way to the thoracic duct). So he tied off these small vessels, and tried the effect of occluding the inferior vena cava. This time there was no increased lymph flow from the thoracic duct, and the lymph protein concentration was lower.

This was real progress. It was clear that: (1) the liver was normally contributing the greater part of the lymph flow to the thoracic duct, and (2) that there was something different about the lymph derived from the liver. Could it have a higher protein content? It did, and it became clear that lymph protein increased when the proportion of liver lymph increased. There was no need to involve any secretory process in lymph formation in this situation; the different permeability of liver and intestinal capillaries to proteins could satisfactorily explain the findings.

Writing about these results in his paper of 1894, he lamented:

Almost in despair, I thought of applying a further test to Heidenhain's view as the source of the lymph obtained under these circumstances [occlusion of the inferior vena cava] . . . I had noticed that the lymphatics in the hilus of the liver were very distended, and it seemed possible that perhaps after all the liver was responsible for the increased flow of concentrated lymph, and that I had during these months been struggling with a chimera. (Starling, 1894a) [The *Shorter Oxford Dictionary* gives three definitions of "chimera": the third one is "a mere wild fantasy; an unfounded conception."]

How could the lowered arterial blood pressure seen in these experiments be associated with increased lymph flow? In terms of the filtration theory this made absolutely no sense. To attack the problem, he brought in his secret weapon, William Bayliss. The two discussed the subject at length. They felt the most likely explanation for the liver producing its increased lymph flow was raised capillary pressure. Heidenhain had noted a fall in arterial pressure when the inferior vena cava was obstructed, and had assumed that this low arterial pressure would give rise to low capillary pressure. Using manometers (Bayliss and Starling, 1894), they carefully measured pressure in the relevant large vessels and showed that the pressure in the portal vein *rose* when the inferior vena cava was obstructed; the liver was congested, and the greater flow of lymph was clearly associated with high capillary pressure in the liver. This was another blow against the secretory hypothesis. In fact, they confirmed that the pressure in capillaries is a reflection of venous and not arterial pressure. In retrospect this makes good sense, because the arterioles (potentially offering a high resistance) come between arteries and capillaries in the circulation, whereas there is no resistance between capillaries and veins.

One of Heidenhain's original experiments was the demonstration that normal saline, given intravenously, produced a large increase on the flow of lymph (he called the saline a "lymphagogue"). He had proposed this as evidence for the secretory origin of lymph. Bayliss and Starling repeated the experiment while measuring several different venous pressures, and showed that intravenous saline caused very large rises in portal vein and vena cava pressure, and, inevitably, large rises in pressure in the capillaries of the liver, intestines and other tissues. So the large increase in lymph flow could be explained by increased filtration resulting from raised capillary pressure in the liver and intestine. It was not necessary to propose that these capillaries "secreted" anything.

The Marriage of Gertrude Starling and William Bayliss

The Starlings were a striking-looking family, with Ernest and his sister Gertrude probably the most handsome. Gertrude was, in the best Victorian way, quite stunning. No one would include William Bayliss among the world's beautiful people, yet he and Gertrude made a remarkable match, and were married on February 21, 1893. He was 32, she 25. They were married in the Catholic Apostolic Church, Harrow Road, London, and there is evidence that William remained a Catholic for the rest of his life (Anon., 1924).

It seems characteristic of the social structure of the time that a research worker should marry his colleague's sister; social life and work were intertwined in a way that no longer exists in scientific communities. Ernest, of course, had only recently married the widow of a colleague.

In 1895 Moses Bayliss died, and William, his only son, was suddenly a rich man. William and Gertrude inherited St. Cuthberts, and the house became a home-away-from-home for many visitors to University College. In the summer there were parties almost every weekend, with Gertrude a memorable hostess, showing special concern for the wives of visiting academics. There were two tennis courts, and a great deal of music, including E. Starling (baritone), F. Starling (piano), and W. Bayliss (violin). Perhaps mixing work and social life was related to the Victorians' skill at entertaining themselves, and playing tennis or music with a colleague facilitated the delicate business of doing research with them.

Lymph (2)

Ernest's work on lymph (most of which was published in the *Journal of Physiology*) was clearly making an impression, because he was invited by the Royal College of Surgeons to give three lectures on lymph—The Arris and Gale Lectures—and these appeared in the *Lancet* in 1894. Remarkably, they were simultaneously translated into German ("Über die Physiologie der Lymphbildung")

Figure 1-8. Gertrude Bayliss (nee Starling) taken in the late 1890s. A Victorian *carte de visite.* (*Family collection*)

and published in *Wien Med. Blätter* in the same year. The lectures provide us with a review of his work up to that time. The first lecture is mostly concerned with Heidenhain's findings, and Ernest, having confirmed the results, is at this time convinced that some of them can only be explained by a secretory process. For example, there was the lymphagogic effect of injecting peptone, sugar and salt: these often appeared in lymph at concentrations greater than their simultaneous plasma concentrations. At the time, this was taken as evidence of secretion, and Starling could offer no explanation for it. [His first paper on lymph had been, in fact, titled "Contributions to the physiology of lymph secretion" (Starling, 1893). He would not use the word "secretion" in the title of any subsequent paper.]

The second Arris and Gale lecture is concerned with the effect of mechanical factors (such as occluding the inferior vena cava) on lymph formation. Starling reviews his own interpretation of Heidenhain's experiments, especially his and Bayliss's finding that arterial pressure did not represent capillary pressure. Because lymph from different parts of the body contained different concentrations of plasma proteins, Starling proposed a hierarchy of capillary permeability: liver capillaries (most permeable), intestinal capillaries (less permeable), muscle capillaries (least permeable). Modern examination of capillary structure with electron microscopy has subsequently shown how liver capillaries have actual gaps in their walls, intestinal capillaries have small specialized areas of high permeability ("fenestrae") and muscle capillaries have no gaps or windows at all. So the microstructure of these vessels has excellent correlation with their permeability to proteins. Starling has never been given credit for his capillary hierarchy.

The third lecture is concerned with two issues: (1) The possible effect of the nervous system on lymph flow and (2) The mode of action of lymphagogues. Heidenhain claimed that nerve stimulation modified lymph flow. Starling repeated the experiments and showed that any changes produced by nerve stimulation were secondary to changes in capillary pressure. The lymphagogue experiments could also be explained by changes in capillary pressure. All that was left of Heidenhain's secretory hypothesis was the unexplained high concentration of certain substances in lymph after they had been put in the circulation. That was the point at which Starling left-off in the Arris and Gale lectures.

Subsequently, he produced an explanation of the high concentrations in lymph. The passage of lymph from the periphery to the thoracic duct takes a finite time—several minutes. So, he reasoned, simultaneous sampling of lymph and blood is not looking at simultaneous events, for the lymph concentration would have been derived from the plasma concentration of several minutes previously. Starling had removed the last strut of Heidenhain's argument. But he emphasized (most diplomatically) that it was only his conclusions that differed from his German teacher:

> I have devoted so much attention to show that the secretory
> hypothesis of lymph formation is unnecessary, that I think ought
> to emphasize the fact that my experiments are merely a continua-
> tion and not a refutation of those of Heidenhain. Not a single
> experimental result in his paper but I have been able to confirm.
> Indeed to Professor Heidenhain's work and teaching I am
> indebted for all the results that I have succeeded in obtaining.
> (Starling, 1894b)

A lesser person might not have resisted the temptation to gloat. But Starling demonstrates great generosity of spirit by actually thanking his German colleague.

Fluid Movement into the Capillary

In most of his lymph research, Starling had considered only outward move-ment of fluid from the capillary and its role in lymph formation. But when, for example, a strong solution of glucose is put into the circulation, the situ-ation becomes more subtle. First, by expanding the volume of circulating blood, the glucose solution raises the capillary pressure in all tissues, the outward flux of water increases, and lymph flow goes up (the lymphagogic effect). Second, the raised osmotic activity of the blood means that simulta-neously water is drawn *into* the capillaries by osmosis. The reader, quite rea-sonably, might find the concept confusing. Starling writes:

> There is no difficulty in reconciling this apparent contradiction if
> we remember that the transference of fluid in two directions is
> due to distinct physical processes. The transference of water from
> lymph-spaces and tissue-cells to the blood is due to a process of
> osmosis. The increased transudation from the blood into the
> lymph-spaces is occasioned by a process which at any rate is
> analogous to filtration. . . . (Starling, 1894b)

The coexistence of two opposite water-moving forces, one hydrostatic and one osmotic, put a very different complexion on the matter. Starling asked whether substances put into the space around the capillaries (the intersti-tial space) pass into capillaries or lymphatics. He asked the question in anes-thetized dogs by making use of the pleural cavity in the chest (the space between the two layers of pleura that surround the lungs). The pleural cav-ity is lined by blood and lymph vessels. Its contents possess a measurable volume, a great advantage for this research. Would a soluble substance put in the pleural space appear first in the urine (i.e., it had passed into the circulation) or in the thoracic duct lymph (i.e., it had been taken up by the lymphatics)? With his collaborator, A. H. Tubby (Starling and Tubby, 1894), dyes dissolved in normal saline were put into the pleural cavity: they used carmine and methylene blue. Within 5–20 minutes of the injection, the dyes appeared in urine; it took several hours for them to color the thoracic duct lymph. They also tried the same dyes in a similar cavity, the peritoneal cav-ity in the abdomen, and obtained similar results.

So, water-soluble dyes entered the capillaries from the outside with ease. The authors then asked a slightly subtler question—what was the effect of changing the osmotic strength of fluid put into the cavity? They reasoned that a solution that had the same osmotic strength as plasma, such as 0.9% saline, should not be absorbed, for there would be no osmotic gradient for the water to pass down. But, to their surprise, 0.9% saline *was* absorbed, as was a 1.2% solution, although this more slowly. They tried the effect of putting serum in to the cavity, and it was not absorbed. So what was differ-ent about 0.9% saline and the serum with respect to their osmotic activity?

For some reason, Starling and Tubby do not ask this question in their paper. Had Starling known what he was shortly to discover—that plasma proteins were providing an important osmotic component in the blood—everything would have been clear.

Starling Forces and "Starling's Principle"

Between 1893 and 1897 Starling wrote nine papers on lymph and capillary function. They were all done at Guy's, surprisingly, with only one of them having Bayliss's name in the title. In 1896 Starling wrote, on his own, the grand finale to all these publications, a paper unexcitingly titled "On the absorption of fluids from the connective tissue spaces" (Starling, 1896). In it, all the issues from the previous papers come together; and his feeling for the broad sweeping generalization takes hold. By the end of the paper there is a real paradigm shift of the way in which every subsequent researcher in the field would have to think about the circulation. The paper has no conventional structure; experiments and discussion are mixed and follow each other as a stream of consciousness. As a publication, it would be quite unacceptable to a modern editor.

The paper begins by discussing the problems of absorbing isotonic solutions (i.e., 0.9–1.0% sodium chloride) into the circulation from the interstitial space. Could it be absorbed by the lymphatics? He answers "no" to this question, because he has tried tying off the lymphatics in the pleural space experiments: this made no difference to the absorption of isotonic saline from the space. But from another source, by a sleight of hand, he produces some new evidence. When an animal loses blood, its blood volume soon returns to what it was before the bleed. The circulating blood becomes more dilute than it previously was, and Starling argues that the only possible way for this to happen is for tissue fluid (which is isotonic) to enter the blood stream across capillary walls. He notes, in passing, that several contemporary workers had explained the dilution by proposing an increase in lymph production. He disposes of this notion by quoting his own lymph experiments, where blood loss *lowered* capillary pressure and lymph flow. So the only possible source of fluid for the dilution of blood after hemorrhage is absorption via the capillaries.

He then does a most elegant experiment to test the hypothesis. The two hind legs of an anesthetized dog were perfused (separately) with dog's blood. One of the legs was made edematous by injecting 1% sodium chloride subcutaneously, and the other leg acted as a control. By using their artery and vein, each leg was perfused the same number of times (between 15 and 25) and the percentage of cells and solid matter, such as proteins, measured before and after perfusion. In every experiment (each was done eight times) the dilution of the blood perfused through the edematous leg was greater than that perfused through the control leg. Starling wrote: "As the result of these experiments, we may affirm with certainty that isotonic salt solution may be taken up directly by the blood circulating in the blood vessels."

His next question is the nature of this uptake. He proposes that in circumstances involving high interstitial fluid pressure (edema) the capillaries show no tendency to collapse. He shows this by examining tissue histologically that has been rendered edematous by injecting fluid. There is no structural reason why excess fluid in the tissue spaces cannot pass into the capillaries, but what actually causes the absorption? He almost sneaks up on the answer:

"I believe the explanation is to be found in a property on which much stress was laid by the older physiologists, and which they termed the endosmotic equivalent of albumen." (Albumen makes up the greater part of the plasma proteins.) He debates the protein concentration on either side of the capillary wall, and notes that it is 8% inside the capillary and 2–3% on the other side: what is the osmotic pressure of this difference?

He then constructed osmometers (all this in the same paper!) that enabled him to measure the osmotic activity of serum. The osmometers were small glass bells having a mouth covered with a piece of peritoneal membrane. Each bell had two tubes at the other end, one for filling the bell with serum; the other for connection with a mercury manometer. The bell was immersed in 1.03% saline (slightly more osmotically active than serum) and left for several days.

After 3 days the height of the column had stabilized at 30–40 mm Hg; he had demonstrated the movement of water produced by a concentration gradient of plasma proteins. He writes:

> The importance of these measurements lies in the fact that,
> although the osmotic pressure of the proteids [proteins] of the
> plasma is so insignificant, it is an order of magnitude comparable
> to that of the capillary pressures [which he was technically
> incapable of measuring]; and whereas capillary pressure deter-
> mines transduction, the osmotic pressure of the proteids of the
> serum determines absorption. Moreover, if we leave the frictional
> resistance of the capillary wall to the passage of fluid through it
> out of account, the osmotic attraction of the serum for the
> extravascular fluid will be proportional to the force expended in
> the production of the latter, so that, at any given time, there must
> be a balance between the hydrostatic pressure of the blood to the
> surrounding fluids . . . here then we have the balance of neces-
> sary forces to explain the accurate and speedy regulation of the
> quantity of circulating fluid.

There is no actual evidence for plasma proteins exerting the osmotic effect; the explanation is just put to the reader as a reasonable bet. Yet Starling arranged his evidence in such a way that no other explanation was possible. The so-called Starling forces have become a cornerstone in our understanding of the circulation. To be precise, there are four forces necessary for an exact analysis: the hydrostatic and osmotic pressures both inside and outside the capillary.

As blood passes along a capillary, there is a fall in hydrostatic pressure within the capillary. This has tempted subsequent authors (especially the writers of textbooks) to construct an idealized capillary in which fluid leaves by filtration at the higher-pressure end, and enters by absorption at the lower-pressure end. In this way there would be little net transfer of fluid. But, attractive as this idea is, filtration dominates absorption in most capillaries of the body, which in no way modifies the truth of Starling's original insight. Modern developments of Starling's ideas are considered in the bibliography for this chapter (see Levick, 1991, 1996). The relationship between the four Starling forces across a capillary can be expressed mathematically (Kedem and Katchalsky, 1958; see the bibliography for further discussion):

$$Jv/A = Lp \{ [P_c - P_i] - \sigma[\pi_p - \pi_i] \}$$

When the plasma protein paper was published in 1896, Starling was 30. Subsequently, there were occasional ripples of disagreement, but it became clear that the new principle (which became known as "Starling's Principle" or "Starling's Filtration Principle") was relevant to every activity involving the secretion or absorption of fluid in the circulation, and, in particular, the disturbed physiology seen in heart and kidney disease. Joseph Barcroft wrote: "It so completely superseded previous work in this field that it put Starling, in his early thirties, into the very front rank of experimental physiologists" (Barcroft, 1928).

He had arrived.

2

1890–1899

Guy's—More Politics

The success of Starling's research from 1890 to 1896 did not seem to make much of an impression on Guy's Hospital. He had to re-apply annually for his poorly paid job and to be assessed by the group of clinicians who made up the Staff Committee (as revealed by the Guy's Hospital minutes, 1890–99). The finances of Guy's were remarkable, though they were probably characteristic of teaching hospitals at the time; there was a hospital fund, guarded by an all-powerful Treasurer (Mr. Lushington) and salaries were made in units ("shares") of this fund. Salaries always had to have Mr. Lushington's approval. In 1890 a share was worth about a hundred pounds, with this value falling steadily throughout the 1890s. So the economy of Guy's had its own inflation.

Staff appointments were made by the Staff Committee, and some relevant minutes involving Starling, Golding-Bird, Washbourn, and Hopkins are reproduced verbatim here. Words like "lecturer," "demonstrator," "senior," and "junior" are used with a certain abandon. Proposals were sometimes voted on, and sometimes not. The Dean had to arbitrate over the distribution of lectures among the teachers. Departmental heads (who did not exist at this time) would have made such decisions outside the committee.

May 9th 1890:

'That Mr E. H. Starling be re-appointed Senior Demonstrator
in Physiology for a period of 1 year from Oct 1st 1890 with the
remuneration of 1½ shares from the school fund.'

May 14th 1891:

'That Dr Starling be recommended for re-appointment as Senior
Demonstrator in Physiology for a period of one year from October
1st 1891 with the remuneration of 1½ shares per annum.'

'That Dr Washbourn be recommended for re-appointment as
Demonstrator of Bacteriology with remuneration of ½ share
annually from the school fund.'

'That Dr Washbourn and Dr Starling be recommended as colecturers in Physiology with Mr Golding-Bird, sharing between
them in equal parts ¼ of the remuneration at present assigned to
the lecturer in Physiology (4½ shares) [this is rather obscure].

May 20th 1892:

'That Mr F. G. Hopkins be recommended for appointment as
Senior Demonstrator in Physiology.'

LOST. Ayes 5, Noes 6

'That Dr Starling be recommended for the appointment'
[no vote]

Amendment made by Mr Golding-Bird, seconded by Mr Lucas

'That fresh applications be invited for the post of Senior
Demonstrator in Physiology.'

CARRIED. Ayes 9, Noes 3 [is this more Golding-Bird malice?]

June 3rd 1892:

Resolved, upon the motion of Dr Perry (Dean) seconded by Dr
Washbourn, 'That Dr E. H. Starling be re-appointed as Joint
Lecturer in Physiology for a period of 2 years from Oct 1st 1892
(No vote; salary 1½ shares per annum).'

'That Mr F. G. Hopkins be recommended Senior Demonstrator
in Physiology.'

Amendment by Mr Golding-Bird, seconded by Mr Jacobson:

'That Mr A. H. Tubby be recommended for the appointment.'

CARRIED. Ayes 7, Noes 3

[We have previously seen Golding-Bird's unpleasantness to
Starling; but his attitude toward Hopkins here seems positively
vindictive.]

July 21st 1893:

[Golding-Bird resigns from teaching Physiology at this time.]
'It was unanimously agreed to recommend to the treasurer that

the assistant lecturers in Physiology [Washbourn and Starling]
should be appointed Joint Lecturers upon Physiology and that as
remuneration they should receive each *two and a quarter* shares
per annum from the school fund.'

June 4th 1894:

'Dr Washbourn moved, seconded by Dr Lucas, that Dr Starling
be appointed senior demonstrator in Physiology' [he had been
re-appointed Joint Lecturer in Physiology for 2 years from Oct 1st
1892]. 'It was agreed to recommend Dr Starling as Senior
Demonstrator of Physiology for a period of 2 years at the remuneration of *one and a half* shares per annum. [In July 1893,
Washbourn and Starling had been appointed Joint Lecturers to
receive 2¼ shares each!]

[As a postscript]: 'that Mr Hopkins was previously recommended as a demonstrator of Physiology and Chemistry for 2
years (1½ shares).'

Starling and Hopkins were too junior to attend these staff meetings, so
they must have been baffled and frustrated by the committee's carryings-
on. By March 1895, Starling had come to the end of his tether, and he wrote
to the Dean (Starling letter, 1895):

> 107 Clifton Hill NW
> March 16th 1895
>
> Dear Dr Shaw
>
> You will remember that last year I wrote to the staff committee
> urging on them the desirability of uniting the practical with the
> theoretical teaching of Physiology and asking for some definite
> statement of my permanent prospects at Guy's.
>
> After considering this letter it was recommended to the
> Treasurer that I should be appointed senior demonstrator for two
> years—with the burden of a large number of revision classes in
> addition to lectures in Physiology and Histology and the elemen-
> tary and advanced practical classes. The consequence is that at
> Guy's I do not make enough to live on, & have no assurance that
> even my present inadequate income will be continued beyond
> next year. I am so burdened with the drudgery of revision classes
> that I have little time or energy left for original work.
>
> Some recent events have made it extremely important for me
> to know, as soon as possible, whether I am to look on my
> position at Guy's merely as a stepping stone to better posts
> elsewhere, or whether the staff and school will make such
> changes in the department as would provide me with a modest
> but adequate income, with a proper control over *all* the

teaching of physiology and freedom from revision classes for the various pass examinations.

These results might be attained by appointing me sole lecturer on Physiology with charge of the practical classes and of the teaching generally and by the appointment of a demonstrator to look after the revision classes at present taken by the senior demonstrator.

Believe me, Dear Dr Shaw

Yours very faithfully

E. H. Starling

This is clearly an ultimatum, and it seems quite justified. Was the mention of "recent events" the offer from Burdon Sanderson at Oxford? Or had there been something more recent? Starling's claim that "I have little time or energy left for original work" seems odd from a man who, in the previous four years, had written several important research papers and a textbook of physiology. Furthermore, he and Florence had translated (from French), Metchnikoff's "Lessons on the Comparative Physiology of Inflammation," which appeared in 1893. Presumably, over the 1894–95 period his teaching load had become significantly heavier.

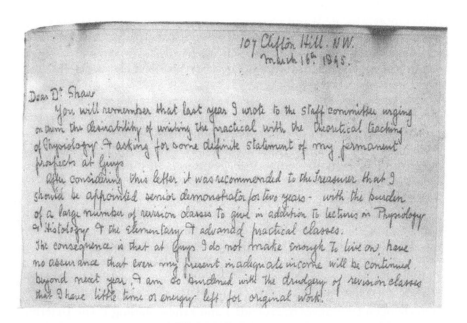

Figure 2-1. Part of Starling's heartfelt ultimatum to the Dean of Guy's Hospital in March, 1895. It seems to have had very little effect on the school's attitude toward him. (*King's College Archives, with permission*)

The staff committee responded to Starling's ultimatum in the only way they knew—they set up a sub-committee. It comes as no surprise that Golding-Bird and Washbourn were both on it, nor is it a surprise that the sub-committee's report avoided answering most of Starling's complaints. The only perceptible improvement was that the total annual salary available for the two lecturers in Physiology was increased from 4½ to 5½ shares per annum. The report did not mention either Starling's name or the possibility of a departmental head.

Meanwhile, Guy's was in the process of building a new medical school. This entry appears in the minutes of the School Committee (Nov 10, 1896):

> Dr Washbourn moved, and Dr Pitt seconded, that a committee be
> appointed to consider what measures, if any, should be taken to
> obtain the requisite funds for the completion of the proposed new
> School buildings and the endowment of chairs of research. After
> considerable discussion, Mr Golding-Bird proposed, and Mr Taylor
> seconded, that the motion be postponed. The amendment was
> lost, and the original proposal adopted by a large majority.

This seems a typically misanthropic gesture by Golding-Bird. In November 1896 the building of the new school must have been well under way, for the enlarged physiology department was opened by the Prince of Wales in May 1897. Why should Golding-Bird have wanted to prevent measures to collect money for it? Was it the suggestion of "endowment of chairs of research" that caused him to react in this way? Incidentally, this is the first mention of endowed chairs in Guy's; it was hardly a new idea, because the chair of physiology at University College, for example, had been endowed over twenty years before, in 1873.

Accounts of the time always emphasize the structural reorganization of the physiology department, though new departments of physiology, anatomy, and pathology were all part of the school rebuilding. The accounts usually ascribe improvements in the physiology department to Starling, though the hospital minutes neither support nor contradict this idea. If he *had* been such a force, it is strange that the 1895 sub-committee did not choose to make him head of the department.

In June 1897, the Physiological Society had their second meeting in Guy's and the minutes record that "The admirable new laboratories were thrown open to inspection" (Physiological Society, Minute books, 1897). The communications included Victor Horsley ("Histology of Red Blood Cells"), Francis Gotch ("Polar Excitation of Nerves with Galvanic Currents"); Gowland Hopkins ("Extraction of Uric Acid After Isolated Meals"; "On Halogen Derivatives of Proteids") and Ernest Starling ("The Determination of the Osmotic Pressure of Colloids" [This was an account of his osmotic experiments with plasma proteins.]) There were fourteen diners and five guests at the dinner—which seems rather a poor turnout for an auspicious occasion. But the Physiological Society minutes of the time show that dinners at London schools often had less than twenty members: it was the meetings at Oxford and Cambridge that really pulled in the diners.

The Physiological Society appointed two secretaries (one Senior and one Junior) for the first time in the 1890s, and Starling and Leonard Hill, from the London Hospital, respectively, held the two posts in 1896. Starling was in fact the senior secretary from 1896 to 1900; in the latter year Hill became senior secretary and William Bayliss junior secretary. In 1903 Bayliss assumed the senior post, and held it until 1922—the longest tenure ever recorded. It meant that, between them, Bayliss and Starling were secretaries at the Society from 1896 to 1922. The senior secretary was—and still is—the chief executive officer of the society.

Meanwhile, poor Hopkins was not getting on too well with the Guy's administration. Like Starling, he was expected to re-apply regularly for his post. In 1897 he seemed to have even more teaching than Starling, for he was a demonstrator in both physiology and chemistry with responsibilities in public health and toxicology. He wrote this *cri de coeur* to the staff committee in June 1897:

> I beg to make application for re-appointment as Demonstrator in Physiology. I would ask at the same time to be excused the duties of demonstrator in Chemistry [here he lists extensive duties of this job and the Public Health Teaching]. I would also like to do research work, and I have found it impossible to do this, not so much on account of the bulk of work which has fallen to my share, but because of the composite nature of my duties, the working day is broken into short sections so as to have no useful leisure. . . .
> I remain, yours very faithfully,
> Fred. G. Hopkins

Hopkins's complaints were not as pointedly expressed as Starling's had been two years before, and the school minutes show no response to his letter. So in 1898 he must have been absolutely delighted to be asked by Michael Foster, the Professor of Physiology at Cambridge, to work there. Foster, characteristically astute, realized that "chemical physiology" was developing into a separate discipline—it was to become biochemistry—and appreciated Hopkins's talent. Hopkins subsequently became the first Professor of Biochemistry at Cambridge, winning the Nobel Prize for his discovery of vitamins in 1929.

Perhaps it is too easy to denigrate Guy's for its treatment of Hopkins and Starling. Money was short, and we have seen that the "share" value fell steadily throughout the decade. Raising one man's salary meant fewer shares for everybody else. Yet, at the same time, there was enough money—presumably from public subscription—to create substantial new school buildings in 1897. We have seen how the first building, the Physiology department, was opened by the Prince of Wales in May 1897 (Guy's Hospital Gazette, 1897). It included a well-designed lecture theatre, seating 400, that has survived the re-organization of recent times. (The theatre has the memorial plaque to Wooldridge on an outer wall.) The next instalment of the school was the Wills library, which was opened in 1903.

The idea of a medical school as a place just for teaching students was steadily disappearing. Even so, by the research standards of the time, Starling's output of a textbook, a book translation and ten groundbreaking papers between 1890 and 1896 was prodigious, and it clearly did not fit the ethos of the medical school. Did sour grapes contribute to the attitude of Golding-Bird and his colleagues? It seems likely, for they were, after all, part-timers confronted by a very gifted professional.

A New World—The Study of Gut Motility

After his memorable paper of 1896 on the absorption of fluids from tissue spaces, Starling wrote only one more publication on capillaries. This was "Glomerular Function of the Kidney" (Starling, 1899). In this research he uses the glomerular capillaries of the kidneys as models for capillaries in which filtration is the only process; for the hydrostatic pressure in the glomerular capillaries is, under normal circumstances, greater than the osmotic pressure of the plasma proteins. When it falls to a pressure lower than the protein's osmotic pressure, urine cannot be formed. This concept grows quite naturally out of Starling's 1896 paper.

Under normal circumstances the nature of the glomerular filtrate is profoundly changed as it passes down the nephron. In particular, it is concentrated by the absorption of water, resulting in urine having a higher osmotic pressure than plasma. Starling's main argument in this paper involves the increase in urine flow (diuresis) resulting from injecting glucose into the circulation. By producing glucose diuresis in anesthetized dogs, he lowered the osmotic pressure of urine to that of plasma. He argues that this can only occur if urine is formed at the glomerulus by filtration alone. It comes as no surprise to learn that Heidenhain had previously proposed an active process to occur at the glomerulus (analogous to his proposed active secretion by capillaries in the formation of lymph). Once again Starling's experiments made Heidenhain's views untenable. Starling actually confirmed this finding in research done twenty years later, when he showed that the activities of the tubules could be paralyzed by cyanide, which rendered urine indistinguishable from glomerular filtrate (Starling and Verney, 1924). We will return to the subject in Chapter 7.

Perhaps he felt that he had had the last word on the filtration/absorption balance in capillaries. In 1936, the distinguished Swedish physiologist August Krogh wrote:

Since Starling's publication [the 1896 paper], the osmometers for
colloids [plasma proteins] have been repeatedly improved and
more accurate determinations of the blood colloids made, but
nothing of a very essential nature has been added to Starling's
explanation. (Krogh, 1936)

So perhaps he stopped at exactly the right moment.

Sometime in 1897, Bayliss and Starling (still working in Guy's; we do not
know whether it was the old or new laboratories) radically changed the direction of their research. They had become interested in the gastrointesti-
nal tract—in particular, gut movements (motility) and their control. Bayliss
had a long-lasting interest in nerves influencing blood-flow (vasomotor
nerves) and it is likely that the pair's interest in the gut stemmed from some
early experiments they did on this concept. They showed in the anesthetized dog that electrical stimulation of the splanchnic (sympathetic) nerves
of the small intestine was associated with a decreased intestinal blood flow
and inhibition of the intestine's rhythmical movements. They measured the
movements by placing a water-filled balloon inside the intestine; the balloon
was connected to a pressure recorder writing on a moving, blackened surface (a kymograph). Thus the size and frequency of the gut movements could
be recorded (Bayliss and Starling, 1899).

Bayliss and Starling soon became interested in the curious gut movements
for their own sake. At the time there was little coherent knowledge of these
movements, and their single serious study begins "On no subject in physiology do we meet with so many discrepancies of fact and opinion as in that of
the physiology of the intestinal movements" (Bayliss and Starling, 1899).
They go on to say that the use of different species, previous feeding, exposure and cooling of the intestines all contribute to the discrepancies in the
literature. By controlling these variables, they managed to produce repro-
ducible and meaningful results, but said:

we must confess that in some instances we have been absolutely
unable to reproduce effects described by physiologists of repute,
however we might vary our method of experiment; and we have
had to come to unsatisfactory conclusions that these results were
due to fallacy of observation or experimental methods.

This might smack of arrogance, were it not that Bayliss and Starling brought
order to a chaotic subject. Their basic experiment involved the anesthetized
dog (which had only been given fluids on the previous day). Gut movements
were recorded either by putting a water-filled balloon in the small intestine
or by a device (invented by Bayliss) known as an "enterograph." This was a
frame holding two needles, whose tips were sewn into the gut wall. One
needle was fixed to the frame, the other pivoted with its upper (mobile)
end attached to a tambour by a thread. Contraction of the muscle of the

gut pulled one needle towards the other and moved the tambour, thus working the recording device. The advantage of the enterograph was that it only detected changes in tension in a single plane, thereby enabling a distinction to be drawn between longitudinal muscle contractions, when the two needles would be placed in the long axis of the gut, and circular muscle, when the two needles would be placed at right angles to the long axis. Such subtleties were not possible using only a water-filled balloon.

Analysis of the movements that Bayliss and Starling saw in the small intestine showed two main types of activity. The first was rhythmical swing-. ing, or rocking, movements that had been named "Pendelbewegungen" (pendular movements) by Ludwig. These occur 10–13 times/minute, and are more marked after food. They originate from muscle itself, and do not depend on the activity of the nerves (Auerbach's plexus) that lie within the muscle of the intestine. This conclusion is based on the lack of effect of nicotine or cocaine on the pendular movement, for these drugs block the activity of the nerves that make up the plexus. The rhythmical pendular movements seem to mix the intestinal contents, for the absorption of the products of digestion depends upon digestive enzymes having access to all the food. This was known before Bayliss and Starling began their work on the subject. Their significant contribution in this field lies in their analysis of the second type of gut movement: peristalsis.

Peristalsis is a more complex type of activity than pendular movement, for it is the process by which the gut contents are actually moved down the intestine. It involves a ring of contraction that usually moves in one direction, being behind the segment of intestinal contents (a "bolus") and gently cajoling it along. The formation of a peristaltic wave depends on the unvarying response of the intestine to local stimulation, and Bayliss and Starling called this "The Law of the Intestine": the law says that local stimulation produces relaxation below the stimulus and contraction above it. In this way a bolus of partially digested food provides a stimulus from inside the intestine that moves the bolus at the speed of the peristaltic wave. They found the phenomenon to be blocked by both nicotine and cocaine, showing that the nerve plexus within the intestine is essential for the transmission of a peristaltic wave. The elegance of the neuronal circuitry is striking, for a region of gut that is in front of a stimulus (usually a bolus) relaxes; as the stimulus passes to this region, relaxation changes to contraction.

Peristalsis is a relatively slow process, a wave taking 1–2 hours to pass from the stomach to the end of the small intestine (some seven meters in man). This is about the time taken for indigestible objects to move this distance, a phenomenon elegantly analyzed by the American physiologist Walter Cannon (1871–1945). Cannon fed cats tinned salmon that contained about 10% bismuth nitrate, a substance that was inert and opaque to x-rays (Cannon, 1902). The cats ate the mixture readily and by using the then recently discovered x-rays ("Röntgen rays" in his article) Cannon was able to follow the radio-opaque food down their digestive tracts. There was no need to anesthetize the animals, since female cats were quite co-operative in the experiments, often curling up and going to sleep between takes. The mixture took about 3 hours to pass by peristalsis from the stomach to the end of the small intestine.

Cannon's main results, published three years after Bayliss and Starling's, confirmed their findings as to the nature of peristalsis. Although Cannon did not contradict Bayliss and Starling on the nature of pendular movements, he greatly modified their descriptions, finding the most conspicuous movement in the cat small intestine to be multiple transverse contractions, which divided the contents into small compartments. He called this "segmentation," noting that it could occur simultaneously with pendular movements. Segmentation, a highly efficient mixing process, recurred constantly as peristalsis moved the contents down the intestine.

There is one respect in which Bayliss and Starling's findings might be criticized. They recorded movements from the empty intestines, mimicking the presence of food by putting in boluses of cotton wool. So they actually studied the movements of the fed intestine; subsequent research has shown different types of phenomena associated with empty intestines ("Housekeeping waves," "migrating complexes," etc.). Early on in their paper they actually note some of these phenomena, but do not include them in the paper's summary, perhaps not fully appreciating the differences between the types of activities associated with fed and fasting states.

Even so, within two or three years, Starling, Bayliss, and Cannon created a whole new paradigm for the nature of gastro-intestinal movements. The findings provided a solid foundation for large areas of modern gastro-enterology and radiology, but for some reason are not often quoted.

The Jodrell Chair of Physiology

Early in 1899, about the time of his 33rd birthday, Starling was elected a fellow of the Royal Society. His twelve proposers included many of the great and the good in English physiology: Schäfer, Sanderson, Sherrington, Haldane, Gaskell, and Halliburton, though Foster was not on the list.

In June of that year, Ernest Schäfer, who had been professor at University College for sixteen years, took the chair of physiology at Edinburgh. This was, on the surface, a curious thing to do, because the Edinburgh department was run down and impoverished; the professor, William Rutherford, had done little research during his tenure. The Cambridge physiologist Walter Gaskell wrote to Schäfer, observing waspishly, "It will be quite a novelty seeing scientific work come out of Edinburgh" (Geison, 1978). Whatever Schäfer's motives for leaving London, his job was advertised, and Starling was one of the applicants; he must still have in high spirits after his election to the Royal Society. The other applicants were William Halliburton, Leonard Hill, George Stewart and Augustus Waller. A selection committee (whose convenor was the surgeon Rickman Godlee—of the "beggarly rudiments") interviewed the candidates (Butler, 1981).

It all seemed straightforward, and on July 28, Godlee announced to the Senate of University College that Dr. Starling had been elected (UCL Council minutes, July 28). There was, however, just one small problem—Godlee himself did not agree with his committee's decision. He had arranged that a Dr. Martin raise an amendment proposing Starling as second applicant— and that Professor Halliburton (one of Starling's proposers for the Royal Society) be elected, "taking into consideration Halliburton's acknowledged superiority as a teacher, his greater experience, the high character of his work, and his general fitness for the post." Halliburton was actually professor at King's College, London, having been there for ten years. He was thirty-nine, six years older than Starling. The Senate then voted on the amendment, which was lost by 10 votes to 8, and Starling had squeaked in. Godlee asked that the names of the voters "for" and "against" should be recorded in the minutes—a manipulative device, surely, for it was asking the voters to line themselves up for or against Godlee himself. One wonders if this device is often used in committees? Is it actually ethical?

Three days later, the Senate's conclusion was presented to the Council of University College (UCL Council minutes, July 31), and the unpleasant ritual re-enacted. Godlee read a statement, not detailed in the minutes, "which he had drawn up with regard to the report and to the position of the Medical Faculty of the Senate." A professor, Sir John Williams, who was not a member of the selection committee and presumably a Godlee plant, proposed an amendment that Professor Halliburton should be elected. A vote followed, and the amendment was lost (no figures were provided). But Godlee, scheming to the end, proposed, in yet another amendment, that the meeting should be adjourned for a week. Voting was five to three against this, and Starling was finally elected as the Jodrell Professor of Physiology. The election was a gruesome slice of university politics, and Starling knew exactly who his enemies were before he had even started the job. Halliburton remained professor at King's College until his retirement in 1928; his textbook of physiology was very successful, and he was a popular lecturer. But his research, by comparison with Starling's, was inconsequential, and it would have been sad for University College if Godlee's machinations had succeeded. They so nearly did. It is of interest that Starling unsuccessfully applied for a chair at another university at about this time (Florence Starling, 1922).

The Jodrell Chair in Physiology had been established in 1873, and Starling was the third holder, being preceded by two of Sharpey's pupils, Burdon Sanderson (1873–1883) and Schäfer (1883–1899). T.J. Phillips Jodrell had been a wealthy and eccentric alumnus of University College (Harte and North, 1991), whose endowment in 1873 was £7,500, and had also started to endow the zoology chair in 1879. Unfortunately, his eccentricity blurred into insanity before he could complete his second gift, and in a remarkable court case (which presumably involved his family trying to claim Jodrell's money) Jodrell was successfully represented by an interesting group known as the Masters in Lunacy. (This body existed to protect the property of lunatics, so that it might be returned to them should they recover. The Masters in Lunacy ceased to exist in the 1930s.) The first Jodrell Professor of Zoology and Comparative Anatomy, Ray Lankester, was then able to take up his post. Lankester was an outstanding teacher, who, incidentally, was instrumental in inspiring the young William Bayliss to become a professional scientist.

The 1890s were clearly an extraordinarily eventful period in Starling's life. We know that three of his four children: Muriel (b. 1893) Phyllis (b. 1894) and John (b. 1898) were born over this period, but sadly no family documents or letters have survived. All that remain are a few snapshots of Ernest playing with the baby girls in a garden. We know little of his feelings or attitudes at the time.

The *Guy's Hospital Gazette*, of August 19, 1899, announced that Starling had been elected to the University College chair. "We congratulate him on his selection for the most coveted physiology appointment in London, but we can only regret the loss to Guy's of one of its best-known men and most original thinkers." There were no expressions of regret in any of the hospital minutes—merely concern that the school would have to find a new teacher. A committee was set up to look into this (it contained Golding-Bird, inevitably), which recommended that the post be advertised. Accounts of Starling's life often claim that he was actually made head of the department at this time. There is no mention of this in any of the school minutes.

Teaching for the winter term was about to begin, and Starling, had he been a different kind of person, would have cast the whole lot adrift. But he didn't; he volunteered to carry on with his own job for as long as necessary. This was an act of extraordinary charity, as his first year was about to begin at University College. Then, in the beginning of the spring term of 1900, the teaching situation at Guy's became even more desperate. The second Boer War (1899–1902) was in progress and several patriots on the Guy's staff temporarily handed in their resignations to work in the Imperial Yeomanry Hospital, South Africa. Among them was Washbourn, and almost unbelievably, Starling offered to do Washbourn's teaching as well throughout 1900. So in the first year of the century, he was professor at University College, while doing his own and Washbourn's teaching at Guy's. It took Guy's a year to fill Starling's post, and in September 1900, Marcus Pembrey (1868–1934) became the first-ever head of the department of physiology.

Starling had once again demonstrated his extraordinary loyalty to the place, having already refused a job at Oxford in 1895 out of consideration to the medical school's feelings (yet Guy's never proposed any memorial to its distinguished son, in contrast to its attitude to Wooldridge). University College might have been rather irritated by this curious person who apparently found it so difficult to tear himself away from his alma mater. Bayliss would have understood, but it is doubtful if the other members of the department would have had enough academic experience to understand Starling's motives.

3

Secretin, Politics, and the New Institute

First Days at University College

When Starling moved to his new department at University College in 1899, it occupied the top floor in the north wing, above the Slade School of Fine Art. (The north wing is on the left side of the quadrangle as one enters from Gower Street.) From its beginnings in 1871, the Slade had been one of the college's most successful departments; within four years it was full to capacity, with 220 students, including many of the college's first women (Harte and North, 1991). In the basement of the north wing was the chemistry department, forming the bottom layer of an interesting academic sandwich.

The physiology department consisted of eleven miscellaneous rooms, including a large one for lectures. The professor's office, wedged between the students' room and the lecture theatre, looks from a contemporary plan to be not much bigger than a cubbyhole. In fact the overall space at University College was less than either the old or the new Guy's laboratories, and we can understand why Bayliss and Starling chose to do their experiments at Guy's. Knowing Starling, he was probably making plans for a bigger department as soon as he had made himself comfortable at Schäfer's old desk.

Apart from William Bayliss, it is not easy to focus on the teachers ("demonstrators") in the department at the turn of the century. There were no

Figure 3-1. The north wing of UCL, photographed in 2003. Its appearance has changed little since it was built in the late nineteenth century. The top floor was the physiology department until 1909, with Starling as professor from 1899–1909. It is likely that secretin was discovered in the large central semi-circular room. The whole building is now the Slade School of Fine Art. (*Author's photograph*)

lecturers, senior lecturers, or readers, for these titles didn't exist (Annual Reports, UCL). The department's only named academic post, apart from the professor, was the Sharpey Scholar: this research scholarship was set up as a memorial to William Sharpey in 1875. In 1899 it was worth £105 a year, and the incumbent usually held it for two years. When Starling was appointed, the Sharpey scholar was Swale Vincent (1868–1934), a physician who had also been "assistant professor" to Schäfer. Vincent was subsequently to play a rather controversial role in endocrinology, denying that many hormones fulfilled all the criteria used in the definition of hormones. His published work at the time was on the adrenal glands and their physiology—he published papers on this topic with Schäfer, whom he rejoined in Edinburgh in 1902.

Another resident was Benjamin Moore (1867–1922) who had been Sharpey Scholar in 1895. He was a chemical physiologist who founded the *Biochemical Journal* in 1906, and after posts at Yale and Liverpool became the first professor of Biochemistry at Oxford in 1920. A second chemical physiologist was William Osborne (1873–1967) who is mentioned in Bayliss and Starling's work on secretin (see below) but left in 1903 to become Professor of Physiology in Melbourne.

There was also a group of clinicians doing physiological research simply out of scientific curiosity. The best-known of these was Sydney Ringer (1835–1910) who was Professor of Clinical Medicine. "Ringers Solution" is clearly a unique contribution to medical science, but the rest of his physiological research looks very like stamp-collecting. Another clinician was Sir John Rose Bradford (1863–1935), a rather more talented physiologist than Ringer; he was working at the time on functions of the kidney that would now be called "endocrine." Bradford had collaborated with Bayliss in the latter's first publication (the electrical changes accompanying secretion) in 1885. Perhaps the most memorable of the part-time physiologists was (Sir) Victor Horsley (1857–1916), pioneer neurosurgeon: he was elected to the Royal Society at 29 and had a laboratory near the physiology department (O'Connor, 1991).

As we shall see, the physiology laboratory moved in 1909 to a new site, and in the intervening years, the Slade School has taken over the top and bottom floors of the building. On the top floor of the Slade School, on the wall of a dim, hardly used corridor, may be found a bronze plaque:

THIS · WAS · THE · DEPARTMENT · OF · PHYSIOLOGY · 1881–1909 ·
HERE · WORKED · SYDNEY · RINGER · J · BURDON · SANDERSON ·
EDWARD · SHARPEY · SCHAFER · JOHN · ROSE · BRADFORD · WILLIAM ·
BAYLISS · AND · ERNEST · HENRY · STARLING

It is an elegant plaque, but nothing seems to be known of its origin; this must be after 1918, for that is the year that Schäfer became Sharpey-Schafer. The sequence of the names corresponds with the year that the individuals first worked in the department.

The period up to the beginning of the 1914–18 war saw a remarkable increase in the number of visiting workers to the physiology department (see Appendix II). They were perhaps attracted partly by what Starling had achieved in the ten years before going to University College, and partly by the experiments of Starling and Bayliss in 1902. These experiments—the discovery of secretin—represent the pinnacle of their joint research.

The Discovery of Secretin

At the turn of the twentieth century, one laboratory, that of Ivan Petrovitch Pavlov, in St. Petersburg, provided a world focus for gastroenterology. Pavlov's school of physiology had elucidated the activities of the stomach, duodenum, and pancreas in digestion. Thus, when the stomach's contents, which were acid, were passed into the duodenum, the duodenum in some way signalled this to the pancreas, and the pancreas secreted pancreatic juice, via the pancreatic duct, into the duodenum. Pancreatic juice was alkaline, and neutralized the acid contents of the gut ("chyme") very precisely. Pancreatic enzymes require a neutral environment to digest food effectively.

Pavlov's school believed that the nervous system was responsible for these phenomena. They argued that when chyme arrived in the duodenum, nerve endings there were stimulated that passed messages to the brain; then, leaving the brain via the vagus nerve, impulses passed to the pancreas and caused the secretion of pancreatic juice. Pavlov had shown that stimulation of the vagus nerve supplying the pancreas could indeed give rise to the secretion of pancreatic juice (though, in retrospect, the effects of stimulating the vagus were never as pronounced as putting acid or food into the duodenum). But research in 1900, by Popielski in Pavlov's laboratory, showed in anesthetized dogs, that when acid was put in the duodenum, pancreatic juice could still be produced. This occurred even when all the nerves passing to the pancreas had been cut. Popielski concluded that the nerves responsible for the secretion of juice must work as local circuits, without involving the brain at all (for he could conceive of no mechanism that did not depend on nerves). Acid put directly into the bloodstream had no effect on pancreatic secretion.

In 1901 two Frenchmen, Wertheimer and Lepage (discussed in Bayliss and Starling, 1902), showed that when acid was put into a piece of small intestine (they used jejunum) separated from the rest of the intestine, and only joined to the body by its blood vessels, pancreatic secretion could *still* be produced by putting acid in to the jejunum. They concluded, like Popielski, that local nerve circuits, acting via the solar plexus, produced this effect. But by not attempting to remove the nerves that they knew accompanied the blood vessels, they made a crucial mistake. They assumed that these nerves were responsible for the effect, so that there was no point in doing the obvious control experiment and removing them.

These results were noted with great interest by Bayliss and Starling, who, as we have seen, had previously investigated the nerve supply to the small intestine. Using anesthetized dogs, they confirmed Popielski's, and Wertheimer and Lepage's, findings: hydrochloric acid put into the duodenum produced, after a latent period of about 2 minutes, a marked flow of pancreatic juice. This effect was still seen after cutting every nerve remotely connected to the pancreas and gut (the vagi, the splanchnic nerves, and removal of the solar plexus). They performed Wertheimer and Lepage's experiment of isolating a segment of jejunum, and then, on January 16, 1902, in University College, did the crucial test. They dissected away the nerves accompanying the vessels to the isolated jejeunum. They then put 10 mL of 0.4% hydrochloric acid into the jejunum, which gave rise to a conspicuous secretion of pancreatic juice (which was recorded by using a drop-recorder writing on a smoked drum). It looked as though nerves from the intestine could not contribute to the secretion of pancreatic juice. The next step was to remove some of the lining of the intestine (mucous membrane), grind it up with hydrochloric acid and sand, filter it, and inject the filtrate into a vein. This produced an even more striking secretion of juice than putting acid into the intestine. It was clear that the acid in the intestine was releasing a substance into blood that circulated and caused the pancreas to secrete juice—a mechanism quite independent of the activity of nerves (Bayliss and Starling, 1902).

It so happened that Charles Martin, Starling's old friend, was visiting the laboratory at University College when the two men were doing the actual experiment. Martin wrote: "a piece of jejunum was tied at both ends, and the nerves supplying it dissected out ... the introduction of some weak hydrochloric acid into it produced as great a secretion of pancreatic juice as had been produced by putting acid into the normal duodenum. I remember Starling saying 'then it must be a chemical reflex.' Rapidly cutting out a further piece of jejunum he rubbed its mucous membrane with sand in weak hydrochloric acid, filtered it, and injected it into the jugular vein of the animal. After a few minutes the pancreas responded by a much greater secretion than had occurred before. It was a great afternoon" (Martin, 1927).

He was right, for the relatively simple experiment represented a great step forward. Yet Wertheimer and Lepage had so very nearly achieved the step: it was just that they were imbued with the Pavlovian idea that everything was explicable in terms of nerves ("nervism"). So, because they were absolutely certain that denervation would prevent the secretion of pancreatic juice, they didn't try denervating the intestinal blood vessels. Bayliss and

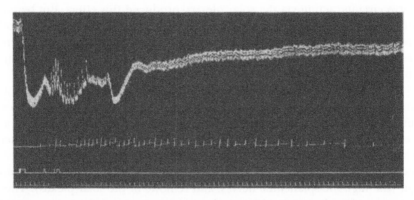

Figure 3-2. The discovery of secretin. An historic kymograph tracing showing the effect of injecting an acid extract of jejunal mucous membrane into a vein in an anaesthetized dog. The lowest tracing is time—the marks are 10 seconds apart. The next tracing records the injection of extracts at time zero, 50 seconds, and 80 seconds. Above this is a recording of drops of pancreatic juice, which operate a drop recorder (the crucial measurement). There is no juice secreted until about 50 seconds after the first injection. The drops increase in frequency until about 2½ minutes, and then slow down. The top tracing is arterial blood pressure. The first injection of jejunal extract produces a severe fall in blood pressure; there is no simple relationship between the subsequent two injections and the falls in blood pressure. These falls are probably the result of vasodilator substances (especially histamine) present in jejunal mucosa. Falls in blood pressure do not cause pancreatic secretion. The experiment is made slightly untidy by Bayliss and Starling making 3 separate injections of jejunal extract. It is not clear why they did this. (*Physiological Society, Blackwells Ltd, with permission*)

Starling, however, had open minds, and they needed to be convinced that everything was controlled by the nervous system. By removing the nerves—and showing that this made no difference—they proved that a totally novel mechanism was at work.

By an odd chance, they had another distinguished visitor when they repeated the experiment. This was Joseph Barcroft (later Professor of Physiology at Cambridge) who was visiting UCL for some other reason:

> Bayliss held a flask in one hand, and the other was introducing a
> tissue extract in to the circulation from a burette. Starling was on
> his haunches, his eye on the level of a cannula which projected
> from the animal: the extract went in: the blood pressure fell for
> the nonce: there was a pause and then—drop, drop, drop, from
> the cannula. There was no secrecy—all was explained without
> reserve, to a man who had published perhaps a couple of papers,
> who hailed from another laboratory . . . and who had no possible
> claim on either the confidence or the genius of these who had
> made so great a discovery. (Franklin, 1953)

Their first publication on the subject (in which they referred to the process as "peripheral reflex secretion," a muddling title) was published a week later (!) in the *Proceedings of the Royal Society* on January 23, 1902. It was here that they proposed the name "secretin" for the substance released into the blood stream. They referred to its stored form as "pro-secretin," a substance releasing secretin under the influence of acid. The experiments were criticized by the distinguished German physiologist Pflüger (discussed in their 1902 publication), who claimed that it was not possible to remove completely tiny nerves from the walls of blood vessels (this is probably true). Starling produced a characteristically robust response: "we submit that since the results of the experiment was such as has been described, it does not in the least matter whether the nerves were all cut or not; the only fact of importance is that it was the *belief* that all the nerves were cut that caused us to try the experiment of making an acid extract of the mucous membrane that led to the discovery of secretin." "It does not in the least matter" is very Starling—direct, and quite close to irritation ("what kind of silly question is that?").

Within a few weeks in 1902, they discovered more about secretin. With William Osborne (a chemical physiologist in the department and mentioned above) they showed that it was not an enzyme, since it resisted boiling water, was soluble in alcohol and was precipitated by tannin. Injection of an acid extract of intestinal mucosa produced a striking fall in blood pressure (see Fig. 3-2).

Bayliss and Starling also showed that the fall in blood pressure was not a property of secretin, but some other substance in the mucosa. They referred to this as "the depressor substance" (it was probably a mixture of

5-hydroxytryptamine, histamine, and related amines). The release of secretin steadily diminished as one moved down the small intestine; this is what would be expected, since the highest concentrations of acid would be nearest the stomach. The "depressor substance" was present uniformly along the intestine.

Pavlov and Secretin

The discovery of secretin had a remarkable effect on Pavlov. In his biography of the Russian physiologist, Boris Babkin describes the day in 1902 when the news of the experiments reached St. Petersburg. Pavlov asked a research worker, V.V. Savich, to repeat Bayliss and Starling's experiment. Babkin writes: "Pavlov and the rest of us watched the experiment in silence. Then, without a word, Pavlov disappeared into his study. He returned half an hour later and said 'Of course, they are right. It is clear that we did not take out an exclusive patent for the discovery of the truth'" (Babkin, 1949).

Pavlov was awarded a Nobel prize in 1904 for his work on the mechanism of digestion, and his Nobel lecture, given in December of that year, involves a detailed discussion of the nervous mechanisms (Nobel lectures, 1967). But the lecture makes absolutely no mention of secretin (which had been described over two years before), or even the possibility than any mechanism other than nerves could be involved. It would seem that if Pavlov hadn't discovered secretin, it was of little interest. Perhaps, after all, he *had* taken out a patent for the discovery of the truth. Pavlov gave up his research on digestion at about this time, though we cannot be sure whether the discovery of secretin was responsible for this. He switched his substantial laboratory resources to the study of conditioned reflexes. It was a subject that brought him far more fame than the mechanisms of digestion, but conditioned reflexes have long since lost much of their interest for scientists.

In spite of their differences, Pavlov did not seem to bear any grudges against Bayliss and Starling. He visited Britain several times, and we have a memorable photo of the guests at a croquet party that includes Pavlov, Starling, and Bayliss, along with their wives (Special Collections Library, UCL). Pavlov would not have played croquet before; his favorite outdoor pursuit according to Babkin, was a Russian game played with two sticks, called "gorodki."

The date of the party is around the turn of the century, which places it uncomfortably close to the discovery of secretin. The site of the party is simply given as "Northwood." One wonders whose house it was. A possible answer is the home of Sir Carey Foster (1835–1919) who became principal of University College shortly after Starling took up the chair of physiology there. Foster's address was a house called "Ladywalk" in Rickmansworth, which is very close to Northwood. I have been to investigate, and found that there is indeed a house called "Ladywalk" on the site, but the original house was demolished in the 1930s and then rebuilt. The distinguished bearded fig-

Figure 3-3. A croquet party at Northwood, Middlesex, given for Pavlov, in about 1902. In the front two rows, from left: Gertrude Bayliss, William Bayliss, J.S. Edkins, Dorothy Wooldridge (with croquet mallet: Florence Starling's daughter by her first marriage), ?Carey Foster (Principal of UCL), unknown woman, Madame Pavlov, Florence Starling, Ernest Starling, Ivan Pavlov, Janet Lane-Claypon, Augustus Waller, unknown man. (*Bayliss Papers, Library Services, University College London, with permission*)

ure (sitting on Madame Pavlov's right) looks as though he is at the center of gravity of the group, and he bears a reasonable resemblance to photos of Foster. But this is only a guess, and sadly, we know the identity of only about a third of the guests in this memorable photograph.

There was an interchange of research workers between University College and St. Petersburg. Thus Boris Babkin (1877–1950) started in Pavlov's laboratory and worked with Starling from 1922 to 1924; he wrote the biography of Pavlov quoted above. Similarly, Gleb Anrep (1891–1955) began in Pavlov's laboratory and worked intermittently at University College between 1912 and 1926. Starling was very fond of Anrep, who learned English well enough to teach in the University College laboratory, before becoming Professor of Physiology at the University of Cairo in 1931. When the Russian government was about to abandon Pavlov in the 1920s, Starling, Anrep, and Bayliss tried, in different ways, to help the Pavlov family (see Chapter 7).

Secretin and the Idea of Hormones

After the discovery of secretin, Starling was invited in 1905 to give the Croonian Lectures of the Royal College of Physicians. There were four lectures, and the first was concerned with the body's control systems; most of these operate through nerves, but there are clearly systems that operate via chemical substances released into blood (secretin represented one such system). In evolutionary terms, this chemical coordination was the older system.

Starling divided the chemical systems of the body into those reacting to external and to internal stimuli. The group of substances responding to internal stimuli "*includes the hormones (from ὁρμάω, hormao, arouse or excite) as we might call them*"[emphasis added]. That is all. This casual reference is apparently the first time that the word "hormone" appears in print. Starling gives no more clues where the word sprang from, though it has been ascribed to W.B. Hardy (a physiologist) and W.T. Vesey (a classical colleague of Hardy's at Caius College, Cambridge) with whom Starling was dining in early 1905 (Needham, 1936). The Greek root "hormao" was not actually new to medical science, for an Oxford physician, John Smith (1630–1679) had described "the hormetick power and contraction of the muscles" in 1666. "Hormetick" in this context seems to carry the notion of "stimulable."

By the introduction of this new word, as well as his experiments, Starling had brought order to a messy subject. The concept of "internal secretion" was already in common use: it was a phrase first used by Claude Bernard in 1855. Bernard used the expression to describe the release of sugar from glycogen stored in the liver, and the contemporary view of internal secretion included any molecule released into the blood stream. Thus carbon dioxide (released from working muscle) stimulated the brain to increase the body's respiration rate: the gas was a perfect example of a messenger substance that was internally secreted. But it was not—to use Starling's new word—a hormone.

The authority on internal secretion was Starling's predecessor at UCL, Edward Schäfer. In 1895, Schäfer gave an influential address to the British Medical Association on the subject, and everything was grist to his mill: he believed that every organ in the body released molecules that influenced other organs. Included in his review was adrenaline, a substance released from the adrenal medulla, which we would nowadays unhesitatingly call a hormone. The recent development of ideas in this field had been very confused, because Schafer's remarks about *all* organs giving rise to internal secretions was very close to the concept known as "organotherapy." This rather disreputable idea had originated some twenty-five years before from Charles Edouard Brown-Séquard (1817–1894).

Brown-Séquard, an Anglo-French physician, made extracts of animal testes and injected the extracts, first into himself, then into patients: he claimed that the procedure brought about rejuvenation. The effect was tacitly assumed to be due to some internal secretion of the testis. We know now that the male hormone produced by the interstitial cells of the testes could possibly have such an effect, but Brown-Séquard had no proper control over his experiments. In an editorial, the *British Medical Journal* wrote in 1889, "The statements he [Brown-Séquard] made—which have unfortunately attracted a good deal of attention—recall the wild imaginings of medieval philosophers in search of an *elixir vitae*" (Anon., 1889). In fairness to Brown-Séquard, it should be pointed out that he also performed a great deal of respectable research, including his classical description of half-transection of the spinal cord (the Brown-Séquard syndrome).

What Brown-Séquard and Schäfer did *not* do was to appreciate the significance of the internally secreting ("ductless") glands. These were lumped in with the catch-all of "internal secretions." Starling realized that there was something unique about the secretions (hormones) of these glands. The secretions were released into the blood in very small quantities, and were very potent, rather in the manner of drugs. He compared their activities to biological catalysts. In his Croonian lectures he made a good case that the pituitary, thyroid, adrenal medulla, endocrine pancreas, and small intestine (giving rise to secretin) all released hormones. There is no mention of sex hormones in his list, in spite of the testicular activity implicit in organotherapy: Starling was presumably not impressed with Brown-Séquard's fantasies.

At about the time of the Croonian lectures, John Edkins (1863–1940), working at St. Bartholomew's Hospital, wondered whether a secretin-like molecule might be in any way responsible for the secretion of hydrochloric acid by the stomach, and Starling discusses this possibility in the third lecture. Following Bayliss's and Starling's experiment, Edkins showed that extracts made from the lining of the stomach (the mucosa), when injected intravenously, cause the stomach itself to secrete acid. This is a rather odd sort of hormonal arrangement, for it seems that the lower end of the stomach (the antrum) releases a hormone that passes into the blood and, after passing round the circulation, stimulates the upper end of the stomach (the fundus) to produce acid. There are very few organs in the body that produce a hormone having an effect on themselves: in this instance the two parts of the stomach are histologically different, and as such are, in a sense, two different organs.

Starling and Edkins both referred to the proposed stomach hormone as "gastric secretin." It took almost 60 years for the substance to gain respectability as "gastrin." It was not until 1964 that Richard Gregory in the University of Liverpool isolated two very similar molecules from the antrum of the stomach; they were different forms of "gastrin." Analysis showed them to consist of chains of 17 amino acids (i.e., they are peptides). At about this time two Swedish chemists, Victor Mutt and J.E. Jorpes, analysed secretin, and showed that it was a chain of 27 amino acids. It is not just by chance that gastrin and secretin were the first peptide hormones to have their structures demonstrated, for they are among the simplest peptide hormones known.

Anti-Vivisectionists: The Brown Dog and Other Matters

Starling's laboratory, as the busiest physiological institution in London, was often under threat from anti-vivisectionists. At the same time, no laboratory had more visits from the animal inspector (established under the vivisection act of 1876). This was not because of any impropriety—it was just that the inspector was the Professor of Anatomy at University College, George (later Sir George) Thane. He visited Starling's laboratory every week or so—he was a conscientious person who was involved with many aspects of UCL. The college's medical library is the "Thane Library" in his memory.

On February 2, 1903, Bayliss and Starling were performing a physiological demonstration in front of an audience of about 60 students (*The Times* law reports 1903). It involved using an anesthetized dog. Bayliss was demonstrating the pressures involved in the secretion of saliva—a remarkable process, because the pressure of the saliva produced by the salivary gland may be higher than the pressure in the artery supplying the gland ("active secretion"). In the audience were two Swedish women, Luid Af-Hageby and Leisa Schartan, who were not medical students, but had special permission to attend. They were actually the joint secretaries of the Anti-Vivisection Society of Sweden. For various reasons, both Starling and Henry Dale (who

Figure 3-4. The "Brown Dog" demonstration. This was a staged version of the actual event. William Bayliss, with an anesthetized dog, addresses a student audience. On Bayliss's right stand Starling, Dale, and Charles Scuffle, who was Starling's technician. (*College Collection Photographs, University College London, with permission*)

was Sharpey Fellow that year, and at the start of his famous career) were involved with the dog. In Starling's words, "It was a small brown mongrel allied to a terrier with short roughish hair, about 14–15 lbs in weight."

The Swedish ladies believed that the dog was conscious for at least a part of the demonstration. They later reported this to the Honourable Stephen Coleridge, Secretary of the British National Anti-Vivisection Society, who drew up a signed statement with the witnesses stating categorically that the dog received no anesthetics. Coleridge read out this statement at a public anti-vivisection meeting at St. James Hall on May 1, 1903, and it was widely reported in the next day's papers. (Coleridge had, in fact, invited the *Daily News* to the meeting.) Had the accusation been true, Bayliss would have committed a criminal offence, and he would have been unfit for his job at University College. Bayliss's son, Leonard, in telling the story more than fifty years later, felt that his father, if left alone, might have done nothing about the libel. Bayliss's solicitor wrote to Coleridge's solicitors, demanding an apology: they received none. So Bayliss's colleagues at University College, especially Starling and Victor Horsley, acted as catalysts, and Horsley is said to have remarked, "The Lord has delivered them into our hands." Persuaded by his friends, Bayliss took out a libel action against Coleridge. Leonard Bayliss felt that it was fortunate that the anti-vivisectionists had attacked his father rather than Starling, for his private income enabled him to face the hazards of a legal action.

The action became known as the "Vivisection Libel Case," or "The Brown Dog Case," and was a *cause célèbre*. It took place at the Old Bailey over four days in November 1903, presided over by the Lord Chief Justice. A queue wound round the court on each day of the trial. The brown dog was actually the subject of an experiment by Starling, who, three months before, had tied off the dog's pancreatic duct. (He did this under anesthetic.) Starling was interested in two conditions of the pancreas: pancreatitis and diabetes mellitus. Did tying off the duct (equivalent to the blockage by a gall-stone) give rise to either of these two conditions? (Bernard, 1856). It seemed unlikely, for in the trial it emerged that the dog had been perfectly well in the intervening period.

On the relevant day (February 1), the dog was first given a morphine injection as premedication and then the standard anesthetic, a mixture of alcohol, chloroform, and ether vapor (ACE). Because it was going to be killed anyway, and to have its pancreas removed and examined, Bayliss used the dog to demonstrate to the class some phenomena associated with salivary secretion. It was during this demonstration that Miss Af-Hageby and Miss Schartan claimed that the dog was showing signs of distress, having not, they claimed, been given any anesthetic. At the end of the demonstration, the dog was taken away by Henry Dale, who killed it while it was still anesthetized, and removed its pancreas for microscopical examination.

The trial was basically the evidence of the two Swedish ladies against the evidence of about a dozen other people who were part of the audience on that day. The first day of the trial included an opening statement by Bayliss's lawyer, Rufus Isaacs KC, in which he claimed that the Swedish ladies were mistaken in their view concerning the dog. He also called one witness, George Woodford, a medical student from St. Bartholomew's Hospital. Woodford stated that the dog was unconscious and showed no signs of movement. Coleridge's counsel, Lawson Walton, KC, then questioned the two Swedish witnesses. They reported what they had said in their original statements to Coleridge, with some additional claims that the dog was struggling as it was brought into the lecture theatre.

The second day began with Bayliss being cross-questioned by both counsels. In reply to Isaacs, he said that he had received no comments about the demonstration—either on that day, or in the period February to May, when the public meeting occurred. In particular, he had received no comment from Coleridge, who had not contacted him before the public meeting on May 1. Under cross-examinination by Walton, Bayliss proposed the view that physiology was the science of living things, and could be best taught by using living animals as subjects. In this instance he wanted the students to witness the fact that saliva could be secreted at a pressure greater than that of the blood supplying the gland—the process of active secretion. Interestingly, Walton spent more time over the necessity of using animals for teaching purposes than the relevant issues of whether this particular animal was conscious. Bayliss confirmed that the dog was deeply unconscious throughout.

Starling was then questioned by Walton, and explained his reasons for tying off the pancreatic duct in February. Furthermore, the dog had a twitch of its left side (probably the result of previous distemper) and this twitch was not altered by anesthesia. Several witnesses had noticed this twitch, though the description of it bore no resemblance to the Swedish ladies' descriptions of the dog's "purposive" behavior. Starling's technician, Charles Scuffle, had helped Starling anesthetize the dog, and Scuffle stated that he saw no sign of consciousness throughout the demonstration. Henry Dale also gave evidence, stating that the dog was unconscious throughout, and that he subsequently killed it in this condition. Four students (Miss Claypon, Miss Barker, Miss Lowry, and Mr. Hume) all gave evidence confirming unconsciousness. Victor Horsley (from UCH) and Professor J.N. Langley (from Cambridge) were not present at the original demonstration, but appeared in court to be cross-examined by Isaacs; they both supported the necessity of using live animals for teaching students. (Such a view would find little support in contemporary medical schools.) Although the issue of the trial was not the use of anesthetized animals for teaching purposes, Isaacs produced these witnesses to ensure the basic justification for using animals in this way—perhaps this was an insurance policy against the jury being diverted by a different issue.

On the third day of the trial, Mr. Walton questioned Miss Af-Hageby, who repeated her assertions that the dog was not anaesthetized; furthermore, she had described this in detail in a book written by Miss Af-Hageby and

Miss Schartan (*The Shambles of Science*). Mr. Walton read out several sections of the book, including a passage in which an unanesthetized dog produced "hilarity" in the lecture theatre: the whole process she saw as being "fun" to the audience. Miss Schartan's evidence was similar to Miss Af-Hageby's.

The fourth day began with the cross examination of the Honorable Stephen Coleridge, who explained that he was a practicing barrister and had been the secretary of the Anti-Vivisection Society for seven years. He had tried to persuade the ladies not to publish their book before the trial, but he had failed. He believed totally in the truth of their statements. Coleridge was

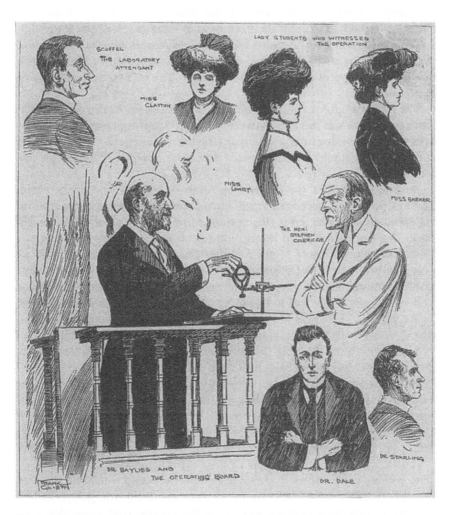

Figure 3-5. Court-room sketches of some of the characters in the Brown Dog trial, printed in the *Daily Graphic*. They are by Frank Gillett, a well-known illustrator of the time. (*Bayliss Papers, Library Services, University College London, with permission*)

interrupted on three occasions by the Lord Chief Justice; the court was not a place for anti-vivisection propaganda, said the Judge. Coleridge did not communicate with Bayliss before the trial because he (Coleridge) would not have believed Bayliss's denial. (In a *Times* editorial published on the last day of the trial, the paper notes: "The Defendant, when placed in the witness box, did as much damage to his own case as the time at his disposal for the purpose would allow.") In his summing-up, the Lord Chief Justice showed his deep suspicions of the defendant's actions: thus he found it strange that the Swedish ladies saw the thing on February 1, yet did not tell Mr Coleridge till April 14. This was all the more remarkable, because it was the first lecture they attended . . . was it not possible that the operation had such an effect upon them as to color their minds as to what they saw? Mr. Coleridge was entitled to advocate his cause, but he had no right to advocate what came by charging others with a criminal offense. "No-one could bring into public discussion charges of a criminal nature unless he was prepared to establish them in court."

The jury was out for about 20 minutes, and found for Bayliss, with £2,000 damages. There was loud applause in the court. Bayliss presented his damages to the physiology department at University College; a writer in *Punch* at the time suggested that the money should be called the "Coleridge vivisection fund."

Despite the verdict, the Brown Dog Affair dragged on for several years. The anti-vivisectionists constructed a bronze statue of the dog on a drinking fountain, and erected it (for reasons which are not clear) in Battersea Park. The fountain had an inscription which was considered insulting to UCL, and it created riots. Medical students came in large numbers from all over London to destroy the statue, and were greeted by anti-vivisectionists. Confrontation between the groups was only prevented by large numbers of police. A strong fence had to be put up around the statue, with permanent police guards. The whole affair became so expensive for Battersea Council that the statue was finally taken away and broken up (Evans, 1964).

Some time after the trial, an odd incident occurred in the laboratory, described here by Lovatt Evans, who was a junior research worker at the time: "One afternoon at tea, a card was brought in by the diener [servant] Fielder with a grin on his face, and Starling, glaring at it, bristled and went out. On the landing, in morning dress, with silk hat, was the Honorable Stephen Coleridge, who said he wished to see round the laboratories. It was the only time I saw Starling almost incoherent with fury as he ordered him off the premises, with Fielder to see that he went" (Evans, 1964).

Coleridge's intentions on that day were not revealed. The early years of the century saw a steady increase in the activities of the anti-vivisectionists, in spite of their poor performance at the Brown Dog trial. Coleridge's declared aim at the time was to make the 1876 Act more stringent (it is worth noting that significant tightening up of the law did not occur until 1986).

In 1906 a Royal Commission on vivisection was set up, with a view to advising parliament. In December of that year Starling gave evidence for three days before the commission, and the subject was considered of such importance that his evidence was published verbatim in the *British Medical Journal.* His statement provides a didactic summary of the uses and importance of animals in research, while emphasizing the necessity of protecting such animals from pain. William Osler, who was sitting on the commission, wrote to a friend (Harvey Cushing) "Starling's evidence is A-1. He's a cracker-jack" (Cushing, 1940). The cracker-jack's evidence was such that the Royal Commission did not suggest any radical changes to the laws governing the use of animals for research.

Starling as Politician and Iconoclast

We earlier touched on some aspects of Starling's character that we could call outspokenness, or perhaps even impulsiveness. In his everyday dealings with his contemporaries this seemed to be a refreshing attribute, for it made him a powerful friend or enemy on committees. Lovatt Evans (who later became Starling's successor at University College) wrote: "at committees his firm wide-pupilled blue eyes would blaze with quick anger at any sign of duplicity, and his tongue was ready on the instant to say 'but this is a trick'"(Evans, 1964). This is not to say that Starling was short-tempered—it is just that he was a passionate being who seemed to have strong views on many subjects. On many nonphysiological topics his writing positively bristles—not against individuals, but organizations. He saw the country's educational system, the organization of medical education, the Harley Street cabal, and (later on, after the Great War) the government itself, as being particularly dreadful. In retrospect, his views have mostly proved to have been right, and in this he resembles Thomas Huxley, who wrote with passion on educational matters (Starling was a great admirer of Huxley). As we shall soon see, Starling attacked the government in public, something that Huxley was never foolhardy enough to do.

The first published signs of Starling's strong views on education appeared in 1903. It was the custom of the time for the *British Medical Journal* to publish annual speeches that opened the academic year for each of the London Medical schools. Most of these were worthy proclamations of loyalty towards the medical profession or the British Empire. But Starling's 1903 offering to the students and staff of University College is shockingly different. He began by telling his audience about the scientific method in medicine. Every patient is a problem to be studied by exact methods; the difficulty is that few doctors have any interest or skill in science. "Many questions are crying out for solutions and all that is wanted is work in the wards by men trained in laboratory methods, and willing to spend laborious days in their application to the problems of disease." (He is promising no quick fix.) "Under present conditions, a man, when he obtains a position on the hospital staff, engages his consulting rooms at a high rental in some fashionable quarter . . . and spends all his spare time scraping together the means for maintaining the appearance of affluence which is supposed to be a necessary condition of success as a consultant . . . he may now become a West End consultant; he will never add anything to the science of medicine."

Starling saw two causes of this malaise: first, there were no academic ideals in London; second: "our medical corporations have grown up from trade guilds, our medical schools from a system of apprenticeship, and we have not yet been able to wipe off the traces of the trade brush with which they were tainted." What would a new medical student make of such radical talk? It is possible that he might not even have recognized the seriousness of the attack on his teachers and the medical profession.

There was worse to come. Starling went on to declare that the remedy was to adopt the German style of medical education, where medical schools had grown up within universities, and doctors were trained in science departments. Recent advances in medicine and science had nearly all come from Germany. He didn't want to replace the English system with a German one, for that (in his words) would be pouring the baby out with the bathwater. He wanted to set up separate Institutes for advanced medicine, headed by professors, with associated laboratories for physiology, pathology, and medical chemistry. The work of these laboratories would be driven by "the constant arrival from hospital of problems arising from the investigations of disease" (Starling, 1903).

These ideas, which later formed the basis of both postgraduate hospitals and medical and surgical "units" now seem self-evident, but the changes took some twenty years to come about. Meanwhile, the University College staff must have been taken aback by the pro-German sentiments and the threats to their Harley Street world. Starling had many contacts with physiologists in Germany and spoke the language fluently; as war became more likely, attitudes in Britain hardened against him.

The "Concentration" Issue

In 1907 Starling's involvement with medical education saw him elected as the medical representative on the Senate of London University. The burning issue of the time was an old problem reborn: it was a proposed institute for preclinical teaching at South Kensington that would be available for students from all London medical schools except University College and Kings. This proposal resulted from the inefficiency of teaching many subjects to relatively small numbers of students, an idea that had originally been suggested by Thomas Huxley (Huxley, 1893).

The proposal became known as the "concentration issue," and it filled many pages of the *British Medical Journal* and the *Lancet.* A proposal was made in 1901 to establish an Institute of Medical Studies at South Kensington; a valuable site was earmarked, and in 1903 a public appeal for money began. There was already a research laboratory on the site, run by Augustus Waller (who was unpaid, with his laboratory supported by Huntley & Palmer's biscuit money). Starling was an enthusiastic supporter of concentration, and his scheme was that Waller's laboratory could be enlarged, and would provide a research establishment to run alongside the teaching facilities of the new institution. This would provide preclinical medicine to any medical school that needed it. Unfortunately two serious flaws developed in the plan. The first was that the money was not forthcoming: in three years, only £70,000, out of a proposed target of £375,000, had been collected, and of this, £50,000 had been provided from South African diamonds by Alfred Beit. The second flaw appears as pure farce: in 1907 the principal of the University of London, Sir Arthur Rucker, categorically denied that the university had pledged any support for the concentration scheme. This was especially disappointing to smaller medical schools that would have benefited from concentration: St. George's, St. Thomas's, and the Middlesex.

By his vigorous support of the South Kensington Institute, Starling incurred a great deal of bad feeling. The most outspoken criticism came from two anonymous letters to the *British Medical Journal,* one of which (signed "Institutio") quotes the other:

> Professor Starling's contribution to your last issue seems to be particularly regrettable. Apparently a quotation from a private letter reached Professor Starling at third or fourth hand. It is alleged to have been written by a member of the Senate "but Professor Starling states that he does not know the name of the author. Is he sure that he has not been imposed upon and may not the authorship be totally different from that which he assumes?" This obscure, sneaky missive must have irritated Starling enormously. (Anonymous letters would not, of course, be published in a modern journal, but they were surprisingly common in the early years of the century.)

It has taken over 80 years for the concentration proposal to become reality, although the recent reorganization of the London schools is not exactly what was proposed in 1903. Starling envisaged the preclinical departments of UCL and Kings staying as they were, while the departments from all the other schools were to amalgamate in Kensington. This proposal was unfortunately seen by the smaller medical schools as a political device for Starling to increase the power of UCL and Kings, and contributed to the animosity shown against him in the medical press. University College has recently concentrated with the Royal Free and Middlesex Hospital medical school, and Kings has joined up with Guy's and St. Thomas's. St. Georges, and Charing Cross, tiny schools in 1903, are now large, single schools. St. Mary's has been swallowed up by Imperial College, a curious echo of the failed plans proposed for South Kensington in 1903.

The Beit Fellowships: Thomas Lewis

However, some benefit—of a very surprising kind—did emerge from the ashes of the concentration issue. When the scheme was abandoned in 1909, contributions to the fund were returned to their donors. This included £50,000 that had been given by Alfred Beit, who unfortunately had died in 1906. The money reverted to his brother Otto, who used it to provide a memorial for Alfred. Otto Beit wrote: "To give this fund such extent as to make it, as I sincerely hope, thoroughly useful, I have decided to increase the above mentioned sum to £215,000, so as to yield by investment in the trustee stocks an annual income of about £7,500. I desire to name this fund 'The Beit Memorial Fellowship for Medical Research.'" After discussion with Starling and Charles Martin, the fund provided thirty selected Beit Fellows with £250 a year. (These fellowships were primarily for use in London, but this was not essential.) Starling and Martin sat on the board of trustees, which included other luminaries such as Clifford Allbutt and William Osler. Beit fellowships provided young research workers with a leg-up to establish their careers, and the fellowships more than tripled the total research funds that were available at the time. The list of recipients reads like a roll of honor for English medical research. Thus, in 1910 Thomas (later Sir Thomas) Lewis and T. R. Elliot (he was also subsequently knighted) were two of the first Beit Fellows, and made use of their fellowships to work in Starling's new laboratory. Elliot and Lewis shared a flat in Mornington Crescent, Camden Town, at this time.

Thomas Lewis was one of the most distinguished medical researchers ever to come out of University College Hospital, and shared with Starling an enthusiasm for Germany and German science, visiting Berlin for several months in the autumn of 1906 to learn the language. But Lewis's enthusiasm for Germany did not extend to lecturing on how much English medicine could learn from the German variety! It was a fascinating time. Between 1908 and 1913 Lewis wrote like a demon, publishing sixty-five research papers, and, among other things, provided the clinical basis for the new subject of electrocardiography. Also, in 1905, he began a new journal (*Heart*), which included Bayliss and Starling on its editorial board. Towards the end of this period in the same institution, Starling began his classic research on the heart as a pump (see Chapter 4), which did not overlap with Lewis's work on the heart's electrical system. So, sadly, these two remarkable men never published anything together.

Revolution: The University, University College, and Its Medical School

The relationship between University College and the University of London had never been easy. Before the 1890s the University was just an examining board, and the teachers of the Colleges found their courses stifled by examination syllabuses over which they had no control. Two Royal Commissions (known as "Selbourne" and "Gresham," respectively) were necessary to produce the University of London Act of 1898. The act changed the University from being just an examining body to a teaching and examining body: teachers now had control of what they taught. The university was reconstituted so that University College, along with other institutions and medical schools, became "schools" of the University. Another important change that was a product of the Gresham Commission, seems almost superfluous to us now. This was the rise in the status of research in the University. "Any limitation of research to institutions set apart for their purpose would tend to lower the academic character of the schools of the university and the standard of teaching" (from the Gresham report). Whether this declaration was a cause of, or merely a response to, a movement that was already in the air isn't clear; but there is no doubt that research in the medical schools of London moved up a gear or two after 1900.

Another proposal in the Gresham report was concentration, discussed above. Medical schools have always been proud and obstinate creatures, undergoing change reluctantly. The findings of the Gresham Commission were drafted as an Act of Parliament by Richard (later Lord) Haldane along with his great friend Sidney Webb. Sidney's wife Beatrice wrote memorably in her diary "Our Partnership" (Webb, 1948):

> Now it so happened that R. B. Haldane and Sidney [Webb]
> united by friendship, made a good combination for the task they
> undertook: to get carried into law the necessary Bill for the
> reorganisation of the London University. To begin with, they
> were, in their several ways, both entirely free from the subtly
> pervading influence of the Oxford and Cambridge of those days,
> with their standards of expensive living and enjoyable leisure, and
> their assumption of belonging to an aristocracy or governing
> class. Haldane had graduated at Edinburgh and Göttingen,
> among students living sparely in uncomfortable lodgings,
> undistracted by games, who looked forward to no other existence
> than one of strenuous brain-work. He believed intensely in the

university, not only as a place for "great teaching" but also as a
source of inspiration by "great minds," producing, in the choicer
spirits, a systematic devotion to learning and research.

There was serious resistance to the act in the House of Commons, many
of whose members were graduates of Oxford or Cambridge, and who felt
threatened by the increased power that the act would bring to London's
University. The resistance was only overcome by a passionate speech from
Haldane in the House.

University College's original hospital in Gower Street (The North London Hospital) was, by 1896, out of date, and Sir John Maple, a furniture
millionaire, provided £100,000 for its rebuilding. By its completion in 1906,
the hospital had cost £200,000, but Maple didn't complain. The cross-shaped
design of the new hospital was by Alfred Waterhouse (the architect of the
Natural History Museum, London, the front of Balliol College, Oxford, and
many other public buildings) and is actually functional, but spikily eccentric in Waterhouse's gothic style. *The Times* (November 7, 1906), wrote chari-
tably: "The style of the building is a free treatment of Renaissance, and the
material red brick with terra cotta dressings."

Meanwhile, the College was determined to strengthen its relationship with
the University, and to go one stage beyond Gresham. It was proposing to
put its site, buildings, and endowments at the University's complete disposal;
the process was known as "incorporation," and it needed its own act of Parliament. Such a plan would mean that the clinical part of the medical school
and the hospital would be run by the university. The University certainly
didn't wish to be responsible for the hospital—like all teaching hospitals, it
depended on charity. Nor was the University prepared to be responsible for
the clinical medical school. In 1903 two schemes for this reorganization were
produced: one by Starling, which proposed a new corporation to run the
medical school and the hospital as a single body, and another by George
Thane (the Professor of Anatomy) proposing the hospital and medical
school to be run by two separate corporations (Merrington, 1976). Starling's
proposal was accepted. The problem was how to create a separate building
for the new school, for one of the necessary conditions of incorporation of
the college with the university was the provision of a site and funds for such
a building (a "school for final medical studies").

A suitable site in University Street, on the other side of Gower Street, and
close to the partly built new hospital, was already owned by the College, but
was too small. In some remarkable way, Starling, along with William Page
May, a physiologist in his department, discovered that a piece of adjacent
land was also available. Its price was £50,000, and they persuaded an old
alumnus of the college to donate the sum—an enormous amount in 1903.
The donor insisted that his name never be revealed (Merrington, 1976).

So, by 1904, a critical stage was reached; all that was needed for the college
to become incorporated was to find the money for the actual building. At the

time a physician at the hospital, Harold Batty Shaw, had a seriously rich patient, Sir Donald Currie, the founder of the Castle Shipping Company. The nature of Currie's illness isn't known, but he believed that Shaw had saved his life. He asked the physician if there was anything he would like, and Shaw answered that he would like a medical school. Currie obliged (it cost £100,000) and for good measure also presented Batty Shaw with a gold watch. Shaw was a popular figure who was Dean in the early years of the new school.

All the necessary pieces of the jigsaw were now in place, and in 1905 the University College (Transfer) act was given Royal Assent. The act included "a separate body corporate by the name of North London or University College Hospital for the purpose of carrying on the Hospital and the medical school." The new hospital was opened in 1906 and the medical school (the School for Advanced Medical Studies) in the same year. The school was designed by Waterhouse's son, Paul: it is a quirky striped building with an elaborate Gower Street entrance that could hardly be used, for it passes straight into the library. As mentioned above, the school recently amalgamated ("concentrated") with two other schools, so that the building lost its original raison d'être as a clinical medical school. Mysteriously, it is now known as "The Rockefeller Building" and seems rather abandoned (2003).

Situated on the south side of University College's quadrangle, that is, on the opposite side to the Slade School and the physiology department, was University College School, a boys' school which had been there since the College began. When incorporation occurred, the school could no longer be part of the college because it would then be part of the university. So in 1907 it was refounded and rebuilt (from public subscription) in Hampstead, where it has remained. (My interest in Starling is related to my having been to both the school and the medical school.)

Starling's Institute of Physiology

Starling realized that the serendipitous acquisition of the school and its playground was a challenge he couldn't resist. He created a place for an entire new preclinical complex of physiology, anatomy, and pharmacology, beginning—not unnaturally—with physiology. His proposed "Institute" of physiology, a title reminiscent of Scottish and German universities, was to be built on the site of the school playground. The "Institute" title wasn't much liked by the college authorities, but it was clear that Starling had become a highly effective committee man, and he had his way.

However, he had to find the money for the new building. It was designed by the Professor of Architecture at the Slade, F. M. Simpson; help with the details came from Starling and Aders Plimmer, a physiological chemist in his department. The cost was about £15,000, with Plimmer himself providing £2,000, and Starling persuading Ludwig Mond, the industrialist, to provide another £3,000. Somehow the rest of the £15,000 was collected, along

Figure 3-6. Starling's new Institute of Physiology, 1909. (*Records Office, University College London, with permission*)

with £5,000 for the fittings. This seems to be very good value, even by the standards of the time.

In appearance the department has an uncluttered northern European air, perhaps reflecting Starling's teutonic enthusiasms. It certainly has not the pseudo-classical look of the first chemistry (1915) and anatomy (1932) buildings that Simpson later designed for the college. The Physiology Institute was officially opened by the Honorable Richard Haldane, Secretary of State for War, on June 18, 1909. The *British Medical Journal* devoted six pages to the occasion, explaining (among other things) why a physiology department might be opened by a Secretary of State for War:

> He is a man of wide and various intellectual interests . . . his
> brother Dr John Scott Haldane of Oxford, is recognized as a
> leading physiologist, while the late Sir John Burdon-Sanderson
> was his uncle. Moreover, the opening of the new Institute is to
> be made the occasion of a review of the University Contingent
> of the Officers Training Corps (OTC), which Haldane has
> called into being; this corps owes much of its present state of
> efficiency to the patriotic efforts of Professor Starling and the
> other professors, all of whom are enthusiastic in the cause.
> (Anon., 1909).

This seems remarkable. In 1909 the Secretary of State for War was presumably thinking about a possible confrontation with Germany, with the OTC intended for use in such a confrontation. Yet Starling was passion-

Figure 3-7. Starling's laboratory in the new institute, 1909. The black bar on the right is actually a very long loop of blackened kymograph paper, which could be used for hours of recording. The laboratory is now a departmental library and meeting room. (*Wellcome Library, London, with permission*)

ately pro-German (a passion that he was to reverse completely when war broke out). To use the occasion of the opening of his new institute for a display of jingo-ism makes little sense. What were his "patriotic" thoughts about? Perhaps the whole occasion might just have been an expression of national pride, unrelated to Germany. For the Great War was five years away and perhaps it was just a coincidence that Haldane was Secretary of State for War.

Like Starling, Haldane was a great enthusiast for German education, having been to Edinburgh and Göttingen universities. His earlier associations with London were (as we have seen) with Sidney Webb in producing the 1898 Act, whereby the University became a teaching institution. But by 1909 a whole new collection of issues was rising to the surface. They included:

1. The status of Imperial College, Kensington, within the University.
2. The site of a new administrative center for the University: Kensington and Bloomsbury were proposed as alternative sites.
3. The possibility of setting up full-time clinical professorial units in London teaching hospitals.

The University proposed a Royal Commission to look into these thorny problems, only to discover that a commission had already been set up. Furthermore, Haldane had already been appointed chairman by the Prime Minister himself. No wonder Ray Lankester wrote, in 1912, "London University . . . the largest body of Committees and sub-committees in the world—elected chiefly by the managing committees of a number of struggling schools and underpaid colleges in London, and so organised as to defeat each other's purposes."

The wide-ranging Haldane Commission was established in 1909 and produced its first report in 1913; Starling gave a great deal of evidence to it. We will return to the commission when we have caught up with some of the discoveries emerging from his laboratory at this time.

4

Starling's Law and Related Matters

The great Dr Starling, in his Law of the Heart,
Said the output was greater if, right at the start,
The cardiac fibres were stretched a bit more,
So their force of contraction would be more than before.
Thus the larger the volume in diastole
The greater the output was likely to be
                                        —A.C. Burton (1972)

The heart is a clever pump. One of its tricks is to pump out exactly the amount of blood that it has just received from the veins. It manages this by automatically adjusting its volume and the force of its contraction to the recently arrived blood. The concept is actually quite an old one in the history of the circulation, but it was Starling who teased out the component parts and was confident enough to dignify his conclusions with a title: "The Law of the Heart" (capital letters seem to be optional). He isolated the organ in his heart–lung preparation, an arrangement that excluded the nervous and circulatory factors that would complicate measurements made in the whole animal.

Experiments before Starling

The germ of the idea can be traced back at least 150 years before Starling. The Reverend Stephen Hales, the Perpetual Curate of Teddington, Middlesex, noting in 1740 the effect of blood loss on the pulse and blood pressure of a mare, found that ". . . violent straining to get loose caused more blood to return to the animal's heart . . . Which must therefore throw out more with each pulsation" (Hales, 1740). Distinguished textbook writers of the time, such as Albrecht von Haller and Johannes Müller, each proposed their own version. In the mid-eighteenth century, Haller thought that blood itself acted as an irritable force when it entered a cardiac chamber. "It is therefore evident" he wrote, "that the heart, stimulated by the impulse of the venal blood, without other assistance, contracts itself." Haller showed that all sorts of stimuli—flatus, watery liquors, wax, or blood—caused the heart to contract more or less forcefully (Haller, 1754). Müller's textbook of 1844 noted that the empty heart could still beat, though rather feebly; the introduction of blood into the chambers caused the organ to contract more forcefully (Chapman and Mitchell, 1965).

These ideas probably came to England via William Sharpey, who, as we have seen earlier, became Professor of Anatomy and Physiology at University College London in 1836. Among Sharpey's many pupils at UCL was Henry Newell Martin (1848–1896). Martin, an Irishman, became the first Professor of Biology at Johns Hopkins University at age 28. (A founding member of both the British and American Physiological societies, he died of alcoholism at age 48.) Martin realized that in order to examine the pumping action of the mammalian heart, the heart needed to be supplied with oxygenated blood—what could be better than using the animal's own lungs for the purpose? (Martin, 1881–82). Furthermore, realizing that he needed to exclude all the nervous and circulating factors that would influence the heart in the normal body, he tied off the blood supply to the rest of the body (including the brain) and left just the key pair of organs—the heart and lungs. The lungs were ventilated with a pump; spontaneous breathing fails in the absence of a brain. The animal (he used dogs) was kept warm in a heated box (see Fig. 4-1). Under these conditions, the heart would beat for several hours, enabling Martin to investigate its intrinsic properties as a pump. He appreciated the importance of maintaining an arterial blood pressure that was similar to that of a normal animal, for only then would the coronary arteries be provided with blood and supply the heart muscle with oxygen. The heart of the preparation beat faster (120–150 beats per minute) than in a normal animal (about 80 beats per minute), because in life, the vagus nerve acts as a brake on the resting heart rate. Exclusion of the nervous system speeds up the heart. Martin found that the organ could maintain an arterial pressure of 86–91mmHg (normal for the dog) at a pulse rate of 118–120/minute. Martin's experiments anticipate Starling's more famous experiments—some 30 years later—quite strikingly. But Martin drew few conclusions from his observations.

Two of Martin's protégés at Johns Hopkins, William Howell and Frank Donaldson, extended his findings by using his experimental arrangement (Howell and Donaldson,1884). In a notable paper they first showed that the volume ejected by the heart (i.e., the left ventricle) in one contraction was between 5 and 10 ml. Raising the arterial pressure by lifting the outflow tube higher above the heart (they used pressures between 58 and 147 mmHg) made no difference to the frequency of the heart or to the volume expelled

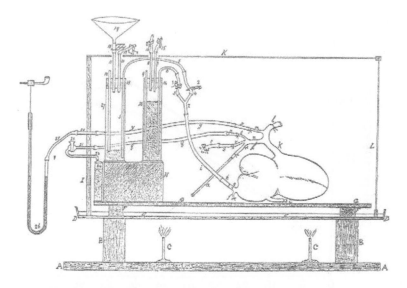

Figure 4-1. Henry Newell Martin's highly diagrammatic version of his heart-lung preparation (1881–82). An anesthetized dog rested inside a warmed box (KL); only the heart is shown. Defibrinated (i.e., whipped) blood was infused into the superior vena cava, via cannula h. The pulmonary circulation is excluded from the diagram. (*Studies Biol Lab Johns Hopkins Univ, 2:119–130, 1881-2*)

per beat. This was a remarkable finding. They proposed correctly that the heart was doing more work sustaining its output in the face of increased resistance. Their last observations examined the effect of changing the venous pressure (i.e., the input pressure) on the cardiac output and the work done by the heart. It is clear from their results that the venous pressure directly controls both the output and the work done by the heart. The results from two of their tables are plotted as a graph in Figure 4-2. Had Howell and Donaldson presented their findings in this way they might have reached the conclusions that Starling was to make some 40 years later. It is remarkable how, in the nineteenth and early twentieth centuries, biological scientists were curiously reluctant to show their results as graphs or to summarize their ideas with diagrams. Graphs and diagrams sum up results in a way that immediately engages the mind and encourages it to draw conclusions. In this particular instance, the drawn lines tacitly ask what happens next: would the heart's output actually fall with increase in pressure? Perhaps the descriptive representation of data was a holdover from the time when biology was natural history.

Meanwhile, the fundamental properties of the heart as a pump were being pursued from other directions. Charles Roy (1854–1899) who became the first professor of Pathology at Cambridge, invented his own device (a cardiometer) which measured the volume of the frog's heart during contraction

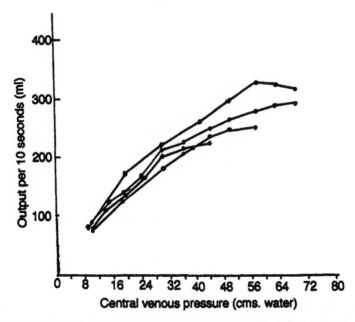

Figure 4-2. Using Martin's heart-lung preparation, Howell and Donaldson (1884) examined the relationship between pressure perfusing the heart (central venous pressure) and the volume of blood ejected per 10 seconds (cardiac output). These four curves have been plotted by the present author from tables in this paper. The values for central venous pressure are inexplicably high. This was probably the first quantitative data ever obtained with a heart–lung preparation. (*Phil Trans 175:139–160, 1884*)

(systole) and relaxation (diastole) (Roy, 1879). His experimental arrangement gave an instantaneous read-off of the beat-to-beat changes in the volume of the heart. He wrote, "At each contraction of the ventricle, in normal circumstances, the quantity of blood thrown out depends upon the degree of distension during diastole" (1879). Roy's papers are largely descriptive and contain few numbers.

The same could not be said for the German physiologist Otto Frank (1865–1944) who spent many years studying the frog heart. Frank was a serious biophysicist: he measured everything—flows, pressures, volumes— that could be measured and recorded data in detail. His most famous work, a 77-page paper "Zur Dynamik des Herzmuskels" (On the dynamics of cardiac muscle) was published in 1895, and has been translated into English (Frank, 1895). Frank provided the isolated frog heart with dilute mammalian blood; it was not a heart–lung preparation. In both Frank's and Roy's frog experiments the hearts were supplied with oxygen via the blood in their ventricles: the gas simply diffused into the muscle without the aid of a coronary circulation. (Only mammals possess such a circulation.) Frank's most relevant experiment studied the tension in the wall of the auricle and ventricle during the heart's activity at different volumes of blood admitted into the respective cavities. It is clear from his findings that the tension (or pressure) generated in the contracting heart was proportional to the volume of fluid in the ventricle. In this ("isovolumic") preparation, blood was not expelled at each beat. This is a different procedure from the one previously described by Martin, and Howell and Donaldson. The full relevance of Frank's contribution will be assessed here after Starling's own experiments have been outlined. The experimental research that led to the heart–lung preparation is summed up in Figure 4-3.

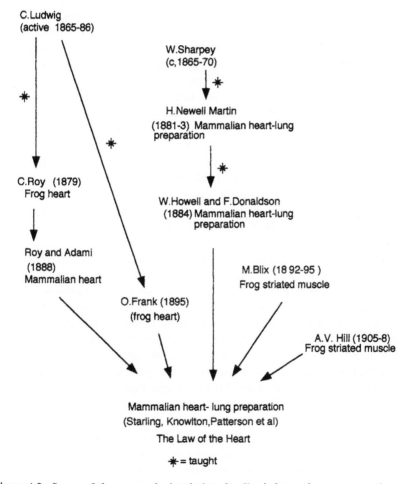

Figure 4-3. Some of the research that led to Starling's heart–lung preparation and the law of the heart. The experiments are outlined in the text.

The Heart–Lung Preparation

In 1906, Starling wrote his last joint paper with William Bayliss. It was a review paper, in German, on hormones (Bayliss and Starling, 1906). Although the two men remained close, their scientific interests moved apart at this time. Starling was planning his new institute and was immersed in university politics, for which Bayliss had little enthusiasm. Starling's research interests were moving inexorably toward the heart; it was not a new subject for him, for he had previously published three lectures on the pathology of heart disease (the Arris and Gale Lectures) in the *Lancet* (1897) while he was at Guy's (Starling, 1897). The lectures not only reviewed a great deal of contemporary knowledge of heart and muscle physiology, but they also anticipated, in a quite remarkable way, the research on the heart that he would do between 1910 and 1914. The first of these lectures is the most relevant, and its theme is the adaptation (or compensation) shown by the normal and the diseased heart. He reviews frog experiments (quoting Roy's work): stretching the ventricle by filling it with saline strengthens its contraction. "This excitatory effect of initial tensions seems to be common to all forms of contractile tissue, and was in fact studied on heart muscle before it had been established for skeletal muscle." He discussed dilatation of the heart in detail: the phenomenon may be physiological—produced at each contraction of the heart—either by increasing the inflow of blood during diastole (increased venous return) or by the heart pumping against an increased resistance. Included in the phenomenon may be the pathological dilation seen in the failing heart. Starling's passion for thinking physiologically about clinical subjects is never more clearly shown than in these lectures. Moreover, he comes within a hair's breadth of stating his Law of the Heart.

He was to publish four papers and two review lectures before his "law" attained full stature. The reader may have a strong sense of *déja vu* when reading summaries of these publications, for many of the findings in them had been made before, though often not in the mammalian heart. The importance of Starling's achievement lay in his ability to bring all the strands together, which he does in a masterly way in the two review lectures.

The heart–lung preparation—a considerable improvement on Martin's original version—was achieved with five of the guest workers in his institute (Knowlton, Anrep, Markwalder, Patterson, and Piper) and is shown in Figure 4-4 (Starling, 1930). It is not quite as daunting as it seems at first sight.

> The circuit can be most easily followed if we begin in the left
> ventricle of the heart (L.V.). Contraction of the ventricle pumps
> blood into the aorta (Ao) which normally conducts oxygenated
> blood to the whole body. In this preparation, all the branches
> except two are tied off. The two remaining are (1) coronary
> arteries, which supply blood to the walls of the right and left

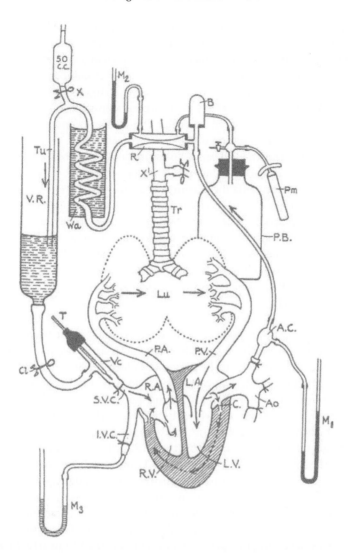

Figure 4-4. One of several versions produced by Starling of his heart–lung preparation. This version was in his textbook *The Principles of Human Physiology* (fifth ed., 1930). See text for explanation.

ventricles (RV and LV) and (2) the innominate artery, which has a glass cannula (AC), tied into it. This provides the output side of the circuit, whose pressure is measured by a mercury manometer (M_1). The blood passes to a variable resistor, an elegant device invented by Starling. This consists of a rubber finger-stall surrounded by an air compartment whose pressure is controlled by a pressure bottle (PB). The higher the pressure within the bottle,

the greater the resistance in the resistor. Most of Starling's
experiments were done with the pressure controlled at around
100 mmHg, and the resistor enabled the effect of changing
arterial resistance on the heart's performance to be examined.
Blood then passed to a warmer (Wa) and then to a large venous
reservoir (VR) whose height above the heart was adjustable. A
wide rubber tube, fitted with an adjustable clip (CI) and containing a thermometer then returned blood to the heart. The
reservoir and the clip enabled venous pressure and flow to be
controlled, though clearly not independently of each other.
Blood entered the heart via the superior vena cava (SVC) and
passed into the right auricle (RA). Pressure on the venous side of
the heart–lung preparation was measured by a water manometer
(M_3). Blood was pumped by the right auricle into the right
ventricle, then to the lungs (Lu) via the pulmonary artery (P.A.),
completing the circuit on the left side of the heart via the
pulmonary vein (P.V.). The lungs were ventilated through a
tracheal tube (X1).

It would not have been possible to circulate normal canine blood through
the circuit, for it would have quickly clotted. An anticoagulant (heparin,
known as "hirudin" at that time) prevented clotting. Starling's source for
this leech extract is not known, but by 1914 he had run out of his supplies,
and had to buy it from his colleague August Krogh in Denmark. He bought
5 g of hirudin from Krogh, for which he paid £13 (Krogh, 1915).

The first of the four "Law of the Heart" papers was written in 1912 with
an American visitor, Franklin Knowlton (1875–1963), who subsequently
became professor of Physiology at Syracuse University, New York (Knowlton
and Starling, 1912). The authors begin by acknowledging their debt to
Martin for his heart–lung preparation. The main improvements in the Starling version were the smaller volume of (dog) blood in the circuit, the resistor to control pressure on the arterial side, and that they examined the
consequences of varying this pressure between 54 and 140mmHg. Provided
that the venous inflow was kept constant, changing the arterial pressure
made no difference either to the rate of the heart, or its output—only to
the work done per beat. This was calculated from the product of volume
per minute and the mean arterial pressure divided by time. When the arterial pressure was raised with the resistor and held at a new level, the heart
volumes—in systole and diastole—increased, only to diminish toward the
previous systolic and diastolic volumes within 20–30 beats.

These phenomena intrigued Starling. His explanation was that the increased arterial pressure first stretched and enlarged the heart's cavity; at
the same time the raised blood pressure increased the coronary blood flow,
providing more oxygen to the muscle, enabling it to contract more forcefully and revert to its original size. He referred to the temporarily enlarged

heart as having poor "tone"; the tone "improved" as the heart recovered its original size.

The only way that Knowlton and Starling could change the actual heart rate in the heart–lung preparation was by altering the blood's temperature. They showed that at temperatures between 25°C and 40°C the heart rate increased linearly from 75 to 170 beats/minute.

They also increased the venous inflow, and found that up to a certain (unspecified) point the output kept up with the venous inflow, though they include no numbers. There was little new in any of these observations, but they provided Starling with a firm base on which to build the rest of the story.

Tone, Adrenaline, and Gleb Anrep

In 1912, while Starling was engaged in these experiments, a messenger arrived from Pavlov. The messenger was Gleb Vassilevitch von Anrep (1890–1955) who was, at the time, a medical student at St. Petersburg (see p. 162). Pavlov had the highest opinion of Anrep, and sent him to Starling on two errands, both related to Starling and Bayliss's earlier work on secretin. The first was to demonstrate to the Londoners that vagal stimulation in the dog really did cause pancreatic secretion; Bayliss and Starling had always believed that the secretion of pancreatic juice was almost entirely hormone (i.e., secretin) driven, whereas the Russian view was that pancreatic juice was the result of nervous (i.e., vagal) activity. Pavlov had, after all, been awarded a Nobel prize for showing the importance of the nervous system in digestion. It turned out that the difference between the two laboratories rested on the morphine premedication that had been given to the animal by Bayliss and Starling, but not by Pavlov (Gregory, 1977). Morphine antagonizes the effect of vagal stimulation. The second part of Anrep's errand was to learn the details of the isolation of secretin. One of the causes of the Russians' antipathy toward the hormone had been their failure to isolate it. This was easily put right.

But Anrep was much more than a messenger. He saw Starling at work on his heart–lung preparation, and was immediately excited by the technique. Within a month or two he had set up his own preparations at University College, and, at Starling's suggestion, written two papers in the *Journal of Physiology* (Anrep, 1912). The writing in these papers gives no hint that they were written by a Russian with only a few weeks' experience of the country. We should not be surprised, for he was a man of many gifts, who ultimately came to speak six languages fluently. One of the two papers, "On the part played by the suprarenals in the normal vascular reactions of the body," emphasised the importance of adrenaline in the circulation—both in the blood and also released from nerve endings. Anrep showed that the phenomenon of heart muscle that was described by Starling as "tone" was very similar to the effect of adrenaline on the heart's action. Anrep concluded that adrenaline produces "a very marked increase of tone accompanied by diminution of systolic and diastolic volumes." The phenomenon became known in the literature as "the Anrep effect." Later on, when Starling fitted the Law of the Heart into the circulatory changes seen in exercise, adrenaline was shown to play a significant role. Subsequently, the changes in heart muscle brought about by adrenaline became known in the literature as an increase in "contractility." Within a few years, "tone" and the "Anrep effect" became virtually extinct expressions.

In 1912, The Royal Society celebrated the 250th anniversary of the signing of its first charter by Charles II, and Pavlov was invited to the celebration. He was also given an honorary degree at Cambridge University. Here is Archibald (A. V.) Hill's account of the occasion:

> [The students] had heard about Pavlov's famous work on
> digestion; and a few of them came to me a day or two before to
> discuss how they could join in celebrating. The idea emerged that
> they should go to a toy-shop and get a large toy dog and furnish it
> with all the rubber tubes, fistulae and glassware there was room
> for; then give it to him in the senate house. They were good
> chaps. When the ceremony was started, there was the dog,
> hanging on strings between opposite galleries. And when Pavlov,
> having received his degree, was returning to his seat, they let it
> down into his arms....he expressed great pleasure, remarking
> "Why, even the students know about my work," as though that was
> the last thing he would expect. [These days, the story would be
> seen as politically very incorrect.] The dog went back to Russia
> with him and remained in his study till Pavlov died in 1936. It is
> now in the Pavlov Museum in Leningrad. (Hill, 1936)

Anrep's third visit to London, in 1914, was cut short by the outbreak of war. He hurried back to Russia with his brother Boris (who was a famous maker of mosaics), qualified in medicine at St. Petersburg, and then joined the army. After being decorated for outstanding bravery in the war, the Russian Revolution found Anrep joining the White Russians. He was captured, escaped, and returned to London, where he resumed research and teaching in Starling's laboratory. He was elected FRS in 1928, when he was 37. This remarkable man was made professor of Physiology at Cairo in 1930, Arabic being one of the six languages he spoke fluently. In 1936, in a preface to the Lane Lectures of Stanford University, he compares the two inspirations of his life (Anrep, 1936).

> Pavlov's personality and that of my later teacher, Starling, had
> equally dominating influences upon my development as a
> physiologist, and I think this preface is the most fitting place in
> which to express my deep regard for these two gallant men of

science. Strikingly different in their general temperament, they were alike in the completeness with which they experienced the unbounded joy of scientific discovery. The Pavlov of digestion was a physiologist of the old school; the Pavlov of conditioned reflexes one could almost say was a physiologist of the future; and Starling was a physiologist of the transition stage between the old physiology of observation and the present physiology of scientific analysis. I humbly hope that in my work . . . I have been able to live up to the high principles which guided my two teachers.

The Law of the Heart (Continued)

The next paper in the Law of the Heart series was by Starling and Joseph Markwalder. (Markwalder is an obscure figure among Starling's collaborators, for we know nothing of his life.) The paper involves the heart–lung preparation, and represents an analysis of the volume of blood pumped our per unit time—the cardiac output. They assessed the proportion of the cardiac output that supplied the heart muscle itself: the coronary circulation. When blood pressure was raised in the heart–lung circuit (by increasing the pressure within the resistor) cardiac output was unchanged, but coronary flow increased. The authors' explanation for this increase was the greater work done by the muscle at raised pressures. This increased metabolism produced substances—carbon dioxide was probably the most important—that dilated the coronary arteries and allowed greater blood flow. Below a certain (unspecified) blood pressure, coronary blood flow ceased, and the heart stopped. Although relevant to the understanding of the heart–lung preparation, these experiments do not directly contribute to the Law of the Heart (Markwalder and Starling, 1914).

In 1912, Sydney Wentworth Patterson, an Australian, came to work with Starling at the Institute. Patterson had qualified in Medicine at Melbourne in 1904, and in 1910 came to work at UC as one of the first Beit Fellows. He married Starling's favorite daughter, Muriel, in 1919; we will return later to the life of the Pattersons. His name appears with Starling's on two of the main Law of the Heart papers: "On the mechanical factors which determine the output of the ventricles" (Patterson and Starling, 1914) and "The regulation of the heart beat" (Patterson, Piper, and Starling, 1914), both appearing in the *Journal of Physiology*. They are not easy papers to read, the second being particularly rambling and obscure. But they are of such importance in the history of the subject that I will describe them both.

The first paper, on cardiac output, begins with a key topic—the connection between venous inflow, ventricular output, and venous pressure. In the heart–lung preparation, Starling and Patterson changed venous input either by altering its pressure (the height of the reservoir above the heart) or by altering its flow (the setting of a clamp on the inflow tube). Whichever way they modified the input, the heart's output varied directly with venous inflow. So long as the functional condition of the heart remains constant, the amount put out at each beat depends directly on the diastolic filling. If we are to understand the factors responsible for the power of the heart to adapt its output to the inflow, it is necessary to study the relation of the pressures on the venous side of the heart to the volume of inflow. The relation is shown in the curves (see Fig. 4-5) in which the venous pressures (VP) in nine different experiments, as measured on the right side of the heart (i.e., the input side) are plotted against the cardiac output.

The figure is remarkable, partly because of the importance of its subject matter, and partly because the axes are plotted the wrong way round. It is a convention in drawing graphs that the horizontal axis represents the variable over which the experimenter has control (the "independent" variable), while the vertical axis represents the variable ("dependent") that the investigator is investigating. In these experiments, Starling and Patterson controlled the venous pressure (expressed as height of the reservoir above the heart) and measured the heart's output in cc's per 10 seconds. The data have been re-plotted in the conventional way in Figure 4-6. It may seem that of the nine

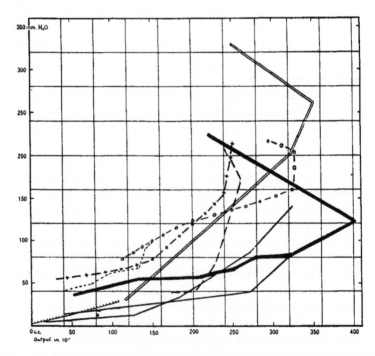

Figure 4-5. The only curves that Starling ever published supporting the law of the heart; there are nine, each from a separate experiment. The axes (venous pressure vertical axis; output horizontal) are the wrong way around (see text). (*Physiological Society; Blackwell Publishing Ltd, with permission*)

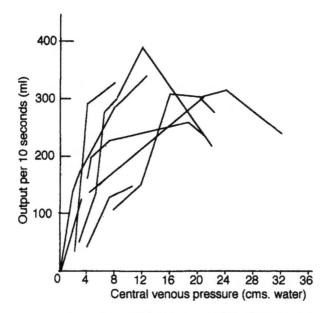

Figure 4-6. The curves from the previous figure, rotated through 90 degrees and replotted. Four of them have a descending limb, to which Starling made no reference, and which have been a source of discussion ever since. But there is no doubt that raising central venous pressure increases cardiac output.

experiments, four reach a peak and have a "descending limb," an observation whose significance was subsequently much debated.

Why did Starling plot this important graph the wrong way round? And why did Langley, the editor of the *Journal of Physiology*, not ask him to change it? The great majority of the illustrations in the journal at that time were smoked drum tracings, or other plots involving time as the independent variable. Under these circumstances there was no uncertainty about the axes. So the explanation may simply have been unfamiliarity. The notion of dependent and independent variables seems to have originated in 1896 from Karl Pearson, Professor of Biometrics at University College, when he used them for his newly-invented calculation of regression (Dr. Eileen Magnello, of the Welcome Institute for the History of Medicine, kindly told me this piece of statistical history.) Perhaps, when Starling plotted these curves, the concept of dependent and independent variables had not yet spread to other branches of science.

One way of summarizing the Patterson and Starling paper is to quote some of its 12 conclusions:

1. The output of the heart is equal to and determined by the amount of blood flowing into the heart, and may be increased or diminished within very wide limits according to the inflow.
2. The maximum output of the heart may amount to as much as three litres per minute for a heart of 56 grms. The maximum

performance of the heart in the heart–lung preparation may
therefore correspond with the maximum output observed by
Krogh in man during severe muscular work. [Krogh had found
variations in the output of the left ventricle in exercising man of
between 2.8 and 21 liters per minute, according to the severity
of the exercise. Starling compared these findings with the dog,
correcting for weight of the hearts.] . . .
6. The greater the arterial resistance, the higher will be the
venous pressure for any given inflow. [This is because the
higher the arterial pressure, the higher the diastolic pressure,
which in turn would be reflected in the input pressure.] . . .
11. With a constant inflow, fatigue of the heart is shown by a rise
of venous pressure accompanied by increased diastolic filling
and mean volume of the heart, the outflow remaining constant.
12. The above statements as to venous pressure apply to both
sides of the heart. When failure occurs under a maximal load,
either the right or the left side of the heart may fail first. [This
has had important consequences in clinical medicine, where
right- and left-sided heart failure are distinguished, and
treated differently.]

These statements seem to say a great deal about the Law of the Heart
without actually stating it: conclusion 1 is close to a paraphrase of the Law.
But it was not until the final paper, "The regulation of the heart beat" by
Patterson, Piper and Starling, published in 1914, that all the constituent parts
fell into place. The new name here is that of Piper (pronounced "Peeper,"
for he was German), who had worked in Rubner's laboratory in Berlin. Piper
was called up for National Service in the German army and was, sadly, killed
on the Eastern Front in 1915.

"The regulation of the heart beat" is 49 pages long and is a lumbering
beast of a paper. It actually occupied a whole edition of the *Journal of Physiology*. It begins in meditative mode. "It is difficult to imagine a more perfectly regulated machine than the heart. In the experiments here described
we have endeavored to determine some underlying principle on which the
heart's power of self-regulation may depend. . . . " But it soon brings us up
short with a discussion of the nature of muscular contraction. Starling is here
very influenced by the work of A. V. Hill and the Swedish physiologist Magnus
Blix. Blix, in a series of publications in the early 1890s, using frog skeletal
(non-cardiac) muscle, had "arrived at the important result that this amount
[the energy set free] was a function of the length of the muscle fibres during the period of contractile stress set up by the excitation." We see here
how Starling is thinking: he is looking for a universal model of muscular
contraction into which cardiac muscle would fit.

At the time Hill was extending Blix's work by examining the heat of
muscular contraction (he usually used the sartorius muscle of the frog), a

technique involving the detection of changes in temperature of around 0.003°C. Hill showed that the energy, measured as heat and produced by contraction, was proportional to the initial length of the muscle. He referred to this as the "chemically active surface of the muscle," a phrase which Starling subsequently adopted. Here is Starling's version of the relationship as given in the paper:

> The law of the heart is therefore the same as that of skeletal muscle, namely that the mechanical energy set free on passage from the resting to the contracted state depends on the area of "chemically active surfaces i.e., on the length of the muscle fibres." This simple formula serves to "explain" the whole behaviour of the isolated mammalian heart,—its movements, powers of adaptation to varying demands made upon it, its behavior in fatigue and under the influence of its nerves or chemical agencies, such as acid ions or adrenaline.

This is the first actual mention of the Law of the Heart. The paper then shows how this law can explain the response of the heart to: (a) raised arterial pressure (this produces no change in cardiac output—just an increase in work done) and (b) increased venous pressure (an increase in cardiac output such that venous pressure tends to revert to normal). Starling adds:

> It is evident that the same reasoning applies to the explanation of the almost instantaneous adaptation of the heart to artificially induced lesions of the valves or artificial stenosis of the aorta. The output of the heart is a function of its filling, the energy of its contraction depends upon the state of dilatation of the heart's cavities.

So the law can offer an explanation of a variety of phenomena seen in the heart–lung preparation.

After these conclusions, this oddly-structured paper then presents us with the actual results from which the conclusions were drawn. The results were obtained using a glass cardiometer (a volume measurer): this contained the two ventricles which were sealed in with a rubber diaphragm, and was connected via a tube to a sensitive piston recorder (a paragraph is devoted to the recorder's extreme sensitivity, for a particle of dust could jeopardize its action.) The recorder made its trace on a smoked drum, and in this way the volume of the heart in systole and in diastole was recorded beat to beat. The records from these experiments are beautifully clear and have often been reproduced in textbooks. But they actually say little that Starling had not said in earlier papers.

Thus Figure 4-7 shows the effect of a rise in venous pressure on the heart's volume. The upper (C) is the cardiometer trace, which is actually upside-down.

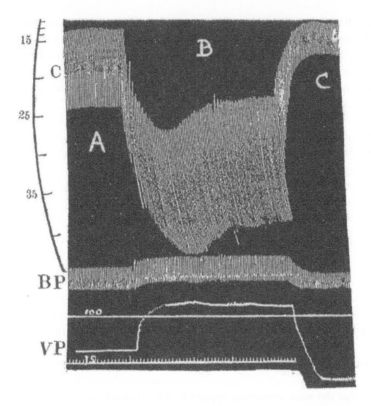

Figure 4-7. One of Starling's most elegant experiments in his papers on the heart–lung preparation. It is a kymograph trace showing the effect of suddenly raising venous pressure (VP) on arterial blood pressure (BP) and on heart volume, measured with a cardiometer (C).The lowest trace is time; with 2-second marks (see text). (*Physiological Society; Blackwell Publishing Ltd, with permission*)

At rest (A) its volume in systole is about 13 ml, and in diastole about 24 ml, meaning that the heart pumps out 11 ml (24 minus 13) per beat. This is twice the output of the left ventricle, because the cardiometer measures the volume of both ventricles. When the venous pressure (VP) is increased (by raising the reservoir in the heart–lung apparatus) the heart is immediately seen to have enlarged. Its smaller (systolic) volume becoming the same as the diastolic volume had been at rest about 25 ml. The larger volume (diastole) now is of the order of 45 ml, meaning that the heart is pumping out 20 ml per beat (i.e., 10 ml per ventricle), keeping up with the increased input that accompanies the raised venous pressure. The period of adjustment that was discussed earlier (first known as "the Anrep effect") lasts for about 30 seconds in this record: the heart then settles down to a new equilibrium until the venous pressure (VP) is changed to a level lower (at "C") than the original baseline, and the heart responds by lowering its output.

The sharp-eyed reader might notice that the original arterial pressure has risen when the venous pressure is elevated: this is the result of squeezing a higher flow through the (unchanged) Starling resistor. If we were to plot a graph of the venous pressure as abscissa against the volume pumped out per beat (ordinate) we would obtain a convex curve that would flatten out at higher venous pressures. Such plots have become known as "Starling curves," and are taken as indices of ventricular performance. The fact that the line slopes at all is an expression of the Law of the Heart: the greater the stretch of the ventricle, the greater the volume of blood expelled per beat. A flatter curve is generally associated with a fatigued heart, and in Starling's experiments was remedied (usually temporarily) by the use of adrenaline.

Fans of medical soaps on television may have noticed certain stereotyped responses when a seriously injured person is rushed into the Emergency room. "I want a venous line in, two units of blood cross matched, and an inotrope." Inotrope is a modern word for adrenaline-like drugs that increase the heart's force of contraction and raise blood pressure, and can be life-saving in a patient who has lost blood. In effect, they elevate the Starling curve; and so, for a given venous pressure, increase the volume of blood expelled per beat. Starling appreciated that there was not just one curve for a given heart, for over the course of a single experiment, the response of heart muscle to increased stretch tended to diminish—an effect which could be partially reversed by adrenaline. If we move away from the heart–lung preparation into the whole animal, we find that different concentrations of circulating agents like adrenaline allow the heart to exhibit a whole range of Starling curves.

The Later History of the Law

Starling's difficult papers in the *Journal of Physiology* would probably have had poor exposure among doctors. But fortunately he was invited to give a prestigious lecture to physicians—the Linacre lecture, which he gave in Cambridge in 1915 (Starling, 1918). This lecture ("The Law of the Heart") wasn't published until 1918, presumably because of the war, but it contained such a lucid summary of his heart–lung experiments that it became one of his most influential works. In it, Starling actually takes a leap into the unknown, for he predicts the behavior of the heart in exercise, a subject upon which he had published nothing. He predicts that the heart would enlarge during exercise, because the exercising muscles would pump more blood into the large veins emptying into the heart, and this extra volume of blood would dilate the ventricles in diastole according to the law. He reckoned without adrenaline, produced during exercise, and we have to wait for the final lecture for this; adrenaline's inotropic effect increases the force of blood expelled at each heart-beat. Furthermore, adrenaline actually increases the heart rate, which contributes importantly to the increased volume of blood pumped around the body during exercise. The heart is often actually *smaller* during exercise as a consequence of this action.

He finishes the Linacre lecture with a swipe at the "so-called practical man." Starling feels that he had been berated for doing research that had no practical value for medicine, for working in an ivory tower. But he was sure that his law had clinical application. He concludes:

> In physiology, as in all other sciences, no discovery is useless, no
> curiosity misplaced or too ambitious, and we may be certain that
> every advance achieved in the quest of pure knowledge will
> sooner or later play its part in the service of man.

His final general statement about the Law of the Heart was a lecture given to the Royal Army Medical Corps in 1920 and published in the Corps' journal (Starling, 1920). This might at first seem an odd outlet for the work, but, as we will shortly see, Starling spent a significant part of the Great War in the RAMC, and had been invited by army colleagues to give the lecture. He once again reviews the conclusions of the heart–lung experiments, and then fits these into the whole animal—in particular, the exercising soldier. He begins this by emphasizing the close parallel that exists in exercise between the volume of blood pumped out per minute (cardiac output) and the oxygen used per minute. The increased oxygen usage produces more carbon dioxide in the metabolizing muscles, which in turn, dilates the small arteries (arterioles) that supply blood to the muscles. This increases the blood flow to the muscles, and, by means of valves in the veins, the increased blood flowing back to the heart is actually pumped there by the contracting muscles. The Law of the Heart then operates, increasing the cardiac output. Adrenaline produced during exercise speeds up the heart and increases the emptying of the ventricles (inotropy; increased contractility). None of this is very different from descriptions that might be found in a modern textbook.

Starling refers to the control that the Law of the Heart confers on the organ's output. He uses the analogy of a motorcycle being driven up a hill. The motorcycle does not slow down, because the rider opens the throttle and lets in more fuel. The heart, when confronted by either increased arterial resistance or by increased flow (venous return) "responds by an increase in the chemical changes of oxidation which, in all human tissues, are the ultimate sources of their available energy." In other words, the heart does more work in response to either challenge. Modern physiologists would see this as a control system: an example of a negative feed-back mechanism. Starling shows us in this lecture that he is aware that a number of other physiological mechanisms may be superimposed on this basic property of heart muscle. In subsequent years physiologists have shown effects (such as the carotid sinus reflex) that have sometimes seemed to contradict the Law of the Heart. But a great deal of research seems to have left the law substantially intact.

Professor Frank and Professor Starling

In 1950, a German cardiologist, Karl Wezler, pointed out that Otto Frank's research on the frog heart had anticipated the results of Starling's heart–lung preparation by about 15 years (Wezler, 1950). Frank's writings—especially his long 1895 paper on the dynamics of heart muscle—are in very difficult German, so it is not surprising that they had escaped many English-speaking readers. (Starling, who spoke German well, had actually discussed Frank's findings in his 1914 paper with Patterson and Piper.) Wezler wrote:

> The so-called "*law of the heart*," which Starling rediscovered ten
> years after Frank [it was actually rather longer], for the warm-
> blooded heart and which is linked to Starling's name, especially
> in the English literature, is implicit in Frank's work on cardiac
> dynamics. One should, more properly speak of the *Frank–Starling
> law of cardiac work* as suggested by Gremels.

Gremels, another German physiologist, had written a paper in 1937 about the action of digitalis on the heart (Gremels, 1937) and it seems probable that this was the first appearance of an association between the names of Frank and Starling. The single relevant sentence in the paper reads: "Unter Zugrundelegung der Frank-Starlingschen gesetze der herzonbeit kann man. . . ." ("on the basis of the Frank–Starling law of cardiac function, one can. . . ."). A strange aspect of the Gremels story is that he actually came to University College and worked with Starling in 1926. The two wrote a paper on the effect of hydrogen ion concentration and anoxaemia on the volume of the heart, using the heart–lung preparation; so Gremels' paper on digitalis glycosides, which used the same preparation, was written 11 years after he had worked with Starling. Why had it taken him so long to attach Frank's name to Starling's law? We are unlikely ever to find out.

The sharing of the credit between the two names is fair. Frank certainly deserves priority for the concept, so it is proper that his name is first. However, Frank was only interested in intrinsic cardiac mechanisms, whereas Starling fitted the relationship into the working of the whole circulation, as we saw in his discussion of exercise. He recognized the phenomenon as part of a self-governing mechanism, as he showed in his motor cycle analogy. But in the literature there is no consistent name of the law—it may be "Starling," "Frank–Starling," a "law," a "mechanism," or a "relationship." Students have even been known to make "Frank" Starling's first name.

The Clinical Relevance of the Law

The law has had important effects on the practice of medicine. Perhaps its most relevant influence is on the study of the disturbed physiology seen in heart failure, for this is associated with a fall in cardiac output, a rise in venous pressure, and an increase in the heart size.

In Patterson and Starling's experiments, two sorts of experimental diminution of cardiac output were seen. The first was produced by increasing venous pressure: the cardiac output (stroke volume) reached a peak (see Figs. 4-5 and 4-6) and then, in four of the curves, fell. This fall has no parallel in the normal activity of the heart; it is possible that under the conditions of the experiment the effect is caused by leakage of the atrio–ventricular valves.

So this particular enlargement of the heart is unlikely to be a useful model for human heart failure. But with the heart–lung preparation a second type of diminution was seen; in the course of an experiment the heart became larger (i.e., its end diastolic volume increased) and the venous pressure increased. Starling referred to this as a "fatigued heart," or one lacking "tone." He showed that it could be "imposed" (i.e., its end-diastolic volume reduced and its venous pressure lowered) with the help of adrenaline. It was tempting to see this "tired" state as a model for cardiac failure.

> Fatigue of the heart may go on to heart failure. This occurs when
> the dilatation, which is the mechanical result of unchanging
> inflow and failing outflow and is the automatic means of regulat-
> ing outflow to inflow, proceeds to such an extent that the tension
> of muscle fibres becomes increasingly inadequate in producing
> rise of intracardiac pressure. The mechanical disadvantage, at
> which in the dilated spherical heart the skein of muscle fibres
> must act, finally smashes up the system and the circulation comes
> to an end. (Patterson, Piper, and Starling, 1914)

The two sorts of dilatation (physiological and pathological) seen in the heart may seem confusing. Arnold Katz has distinguished them thus:

> A "modern" way to resolve this apparent contradiction is to view
> Starling's Law of the Heart as a short-term adaptive functional
> response in which dilatation increases the work of the heart,
> whereas dilatation of the failing heart . . . results from a long-term
> maladaptive architectural response caused by abnormal prolifera-
> tive (transcriptional) signalling. (Katz, 2002)

The notion of heart failure just as a functional deterioration of the heart muscle now seems hopelessly simplistic; endocrine and renal changes have roles in this complex syndrome where cause and effect are weirdly confused.

But none of them alter the basic relationship involving venous return, end-diastolic volume and cardiac output that Starling analyzed.

Since the early twentieth century, hundreds of cardio–vascular investigations have been proposed examining the validity of the law, but it would be misleading to suggest that they all confirm it. The problem is that the experiments make use of many indirect versions of venous pressure and stroke volume. There is no right and wrong. Here is a fearsome description of some of the parameters that have been used plotting Starling curves in humans, provided by Levick:

> The curve appears in many guises. For the abscissa, central
> venous pressure is often chosen because CVP is easily measured
> by catheterization and is an important regulator of average fibre
> length; however its relation to the fibre length is indirect and
> non-linear. Other indices of stretch include right ventricular end
> diastolic pressure, left ventricular end diastolic pressure, ventricu-
> lar end diastolic volume measured by 2-plane cineangiography,
> and radionuclide angiography [and so on]. (Levick, 1991)

It is hardly surprising that some investigators have obtained data that leaves them unconvinced.

There are several circumstances in man when the influence of venous filling pressures on the heart's output can be clearly shown (McMichael, 1950):

1. Posture exerts an important influence, for in the upright position the pressure in the great veins falls. In a resting recumbent adult, the output of the heart may be 5.3 litres per minute, falling to 4 litres per minute on quiet standing.
2. It has been demonstrated by Cournand's catheterization studies of the right side of the heart that the pulse pressure (and therefore the stroke output) varies with each phase of the respiratory cycle.
3. The low cardiac output seen in haemorrhagic shock responds dramatically to intravenous blood (or, for a short time, saline).
4. In some normal subjects, saline infusion (increases venous filling pressure) produces a rise in cardiac output.

But there is one rider to the law that gives it the most resounding scientific and clinical validity. Surprisingly, Starling did not seem to have thought of it, though he wrote of the differing effects of right-sided and left-sided heart failure. In the words of W. F. Hamilton, in 1955:

> I believe that the law does play a very important role in everyday
> life, and that that role is to preserve the balance between the
> pumping of the right and left ventricles. This balance must be

very exact, and, since the two ventricles are subject to the same
hormonal and nervous influences, they each act as a control for
each other. (Hamilton, 1955)

It seems the best possible vindication for the law.

The reader might feel that the Law of the Heart has taken up a dispropor-
tionate amount of space in this life of Starling. He would be right, for as a
scientific achievement it probably ranks below his contributions to the mi-
crocirculation and to endocrinology. Yet it is a concept that draws together
a host of experimental threads, and continues to be relevant in any discussions of the whole circulation, remaining a subject for debate. (How, for
example, can we best assess the length of a cardiac muscle fiber in life?) This
does not seem to be the case with any of Starling's other important contributions to physiology, which somehow lack this contentious edge.

Interlude: The Haldane Commission (1910–13)

Haldane's Royal Commission

Lord Haldane's giant enquiry looked into many aspects of the University of London, but concerned itself especially with the organization of medical education. Starling was probably its most loquacious witness. The Commission's evidence, taken from about a hundred witnesses, fills hundreds of pages (Royal Commission Reports, 1913). Fortunately for us, it has a well-written summary.

This summary reminds us that the reorganization of the university (resulting from the Act of 1898) involved: (1) the establishment of a faculty of medicine, which consisted of appointed teachers of the university; (2) Boards of studies for each subject in the medical curriculum—these boards having the power to design the curricula for their subjects; (3) The twelve London medical schools being admitted as Schools of the University; and (4) Internal students studying at a recognized school or schools for 5½ years, including 3 years of advanced (i.e., clinical) teaching.

"These changes" the summary notes, "do not appear to us to have accomplished very much." The faculty of medicine was probably too big to be of any use—teachers' loyalty remained with their schools. So the schools were as detached from, and as independent of, the University as they were when the University was just an examining board. The summary continues:

"As Professor Starling has pointed out in the very able statement he submitted to us, the main difficulty of creating a University School of Medicine in London arose from the evolution of medical teaching." This was a subject on which Starling had well-formed views (Starling, 1903). The medical schools, his evidence began, were older than the university, and so had preceded any attempt to teach medical sciences: teaching began as apprenticeships to apothecaries (in rural areas) or "walking the wards" (in the teaching hospitals of cities). London medical schools sprang up as appendages of the hospitals, but had no formal existence. Their finances were derived from charity and student fees; the money was pooled and divided into "shares" (as we saw from Starling's experiences at Guy's Hospital in Chapter 2). The shares provided the wretched salaries of preclinical staff, who received no fees from students. Such staff had originated with chemistry teachers, forty years before, that is, in about 1870. Subsequently physics, biology, anatomy, and physiology became part of the syllabus, and the teachers of these subjects had to be paid with shares.

The total cost of the preclinical departments had grown so much that in some schools the shares had actually become minus numbers, with contributions from the clinical teachers having to make up the shortfall. The greater part of the work done in the wards was performed by the clinical students as "clerks" in medical wards and "dressers" in surgical wards. These were the vestiges of the apprentice system. This clearly saved the schools money, and provided practical education for the students. But it meant that schools were often teetering on the edge of bankruptcy, and it was essential to attract enough students. To help balance the books, graduates were expected to send their patients to hospital at their old school. In this way, loyalty to the school and self-interest became inseparable.

So, Starling argued, schools were not really in the interest of the university, or of medicine itself. However efficient a school might be, it had the nature of a "trade school," with the standards set before the student based solely on professional success. Students were not impressed by the teaching of science, done by men—"not of marked ability"—whose only job was to get the students through necessary exams. The idea of the first two or three years was not educational, but professional. Nor was Starling impressed by the teaching of clinicians on the wards; it was characterized by empiricism and ignorance of the science upon which medicine is founded. Starling's views clearly made an impression on the members of the commission, for he was often quoted in their discussions.

Another important witness to the commission was the American educationalist Abraham Flexner (see bibliography), who had published a well-known study of medical education in America and Canada. Flexner's evidence—about a quarter the length of Starling's—was concerned only with the establishment of clinical units. (His evidence was not concerned with the relationships between London medical schools.) He was particularly scathing about English hospital consultants: "It is still supposed that because a man is an accomplished physician he is an excellent teacher. Clinical teaching in London remains an incident in the life of a busy consultant who comes to his post through promotion on the basis of seniority and visits his too miscellaneous wards twice weekly, between perhaps, two and four, in company with his house physicians and clinical clerks . . . there is no interaction between scientists and clinicians . . . there is no reward stimulating the young physician to engage in original work; his rewards are those of faithful routine." Flexner went on to say that university teaching can only be given by men who are actively engaged in the advancement of knowledge in the subject they teach. His remarks had a profound effect on the commission.

Sir William Osler, the Regius Professor of Medicine at Oxford, was another witness. He, like Starling and Flexner, attacked the nonscientific nature of clinical teaching, believing that the way forward was the setting up of clinical units within hospitals, each headed by a professor. "A professor of medicine requires the organization of a hospital unit, if he is to carry out his three-fold duty of curing the sick, studying the problems of disease, and not only training his students in the technique of their art, but giving them university instruction in the science of their profession" [a superhuman task]. The cost of clinical units should be shared between university and hospital. Such arrangements would be similar to the system used in German medical schools. The commission pointed out that not everything was inferior in London medical education: the time-honored system of clerks and dressers in the teaching hospitals gave students valuable experience with patients. American and German students had no such opportunities in their education, although Osler described how he had recently successfully introduced clerks and dressers into the curriculum of Johns Hopkins Medical School. The system subsequently spread to many of the other major medical schools in America.

The Concentration Issue

The commissioners had great problems with this old chestnut. It was their view that medical schools—however small and inefficient—were building more laboratories and gradually acquiring more specialized preclinical teachers. (We saw earlier how Guy's Hospital was enlarging its preclinical faculties just before Starling left in 1899.) At the same time, the commission noted that these jobs were, by university standards, poorly paid, and so would hardly attract the best candidates—a heartfelt point raised by Starling in his evidence. The so-called preliminary sciences (physics, chemistry, and biology), which medical schools found expensive and time-consuming to teach, were gradually moving into the curricula of schools. The commission took evidence from a science master at Harrow School, who observed that many schools were actually building laboratories and improving their teaching of these subjects. The General Medical Council had recently taken to inspecting schools' science departments with this in mind. This teaching in schools was in effect an argument against concentration, for it reduced the need for these preliminary subjects to be taught by the university. Another argument against concentration was that if all preclinical teaching were done in the proposed three large centers (South Kensington, UCL, and Kings) and were not part of a hospital environment, the clinical relevance of preclinical teaching would be lost on the students. In his evidence on concentration, Starling said:

> The drawbacks associated with the existence of so many small and
> inefficient schools of Medicine in London have long been recog-
> nized . . . as a result a scheme was set in foot for the foundation of
> an Institute of Medical Sciences in South Kensington. For the
> history of this scheme I may quote from the BMJ of the 13th April,
> written just before the scheme finally collapsed, and the senate
> decided to give back to the donors the money already collected for
> the purposes of the institute.

In quoting from his BMJ article, he was withering in his contempt for this decision:

> The Medical Faculty approved the scheme *nemine contradicente*; it
> has now waited until the King has given his warm approval, the
> Prince of Wales expressed his fullest sympathies with the proposal
> . . . and now, at the eleventh hour it [the Medical faculty] turns
> round with barely an expression of regret, and with no attempt to
> assign a reason of any weight in excuse for its fickleness, decides
> that a central institute no longer meets with its approval.

The commission failed to find enough evidence to support concentration. The collapse of the scheme some months before had given the commission little choice! It has subsequently taken about 80 years for the medical schools of London to concentrate (as we saw in Chapter 3). But the commission clearly had been influenced by Starling's views on other aspects of medical education, particularly the concept of the clinical unit.

The Next Step

In May 1913, Starling wrote two articles in the *British Medical Journal* summarizing the Royal Commission's findings (Starling, 1913a,b). Perhaps they might have been better received if from a member of the commission itself—Starling was, after all, a witness who had his own axe to grind. He outlined the commission's conclusions, stressing the necessity of University teaching in both preclinical and clinical subjects.

The English schools of physiology and pathology, and of tropical
diseases, have acquired a high reputation in the civilised word,
but reference to the authorities quoted in any textbook of
medicine shows that, with but a few honourable exceptions, the
recent advances in the science of medicine . . . are accredited to
some foreign school.

He sees the most important proposals of the commission as the appointment of university professors in clinical subjects, that is, the establishment
of clinical units. These professors would be in charge of an out-patient department, "and will be expected to devote all their heart and the greater
part of their time to the work of teaching and research of their department."
Some private practice, say two afternoons a week, was suggested, and a salary of £1,000—£1,500 a year thought appropriate. (Starling's annual salary
as Jodrell Professor was £260 at this time, though Lovatt Evans writes that it
was supplemented by £300 from student fees.) Starling goes on: "There is
no question that these proposals will meet with much criticism. Mediocrity
is certain to be up in arms at the institution of any system which promises to
pick out and place in a superior position the talented, the enthusiast, or the
man of genius." He was careful to say that no threat existed to the clerk/
dresser appointments universal in London medical schools, these being seen
as strengths of the London system.

Mediocrity certainly was up in arms over the proposals. A dermatolo-
gist from St. Mary's Hospital, E. Graham Little, replied in a letter to the
British Medical Journal, "Professor Starling is obviously disappointed that
the commissioners have not—ostensibly—revived his pet scheme of concentration" (Little, 1913). Little's comments are essentially Luddite: secondary schools, he asserts, are not suitable places to teach chemistry,
physics, and biology. But it is the subject of clinical units that really stirs
him up. "[The commissioners] ignored the real experts, and have produced practically in its entirety the fantastic scheme of one or two witnesses;
one, the chief and perhaps 'the only begetter,' Mr Abraham Flexner, is
an American layman who exhibited the most ludicrous ignorance of our
English system; the other a physician who had spent all of his active professional life in America." (Little carefully avoids writing the name "Osler,"
as though it were one of the names of God). A list of fourteen physicians
is then produced, all of whom had given evidence *against* clinical units:
"Professor Starling attempts to counter this serious opposition by a mere
flippancy: 'Mediocrity' he says 'is certain to be up in arms at the institution
of any system . . .'" There follows an attack on German medical teaching.
Little claimed that clinical professors in Germany abused their positions
by doing private practice at the expense of research. He thought that this
would inevitably happen in Britain. He complained, in a *reductio ad absurdem*
argument, that the Commissioners wanted to adopt the German system
entirely, which was never their intention.

Neither Starling's nor Little's articles caused any ripples on the waters of the *British Medical Journal,* for the introduction of the changes proposed by the Commission was overtaken by the Great War. Starling was later to be deeply involved with the first clinical units, though perhaps not in a way that the reader might expect. The introduction of his other brain-child—concentration—into London medical schools has come about, but has taken a very long time.

5

The Great War

Gas! Gas! Quick boys!—An ecstasy of fumbling,
Fitting the clumsy helmets just in time:
But someone still was yelling out and stumbling
And floundering like a man in fire or lime . . .
Dim, though the misty panes and thick green light,
As under a green sea, I saw him drowning
      —Wilfred Owen: "Dulce et decorum est"
           (from "Collected Poems")

Military Physician at Woolwich

Archduke Ferdinand was assassinated in Sarajevo at the end of June, 1914, and Britain declared war on Germany on August 4. At that time the Starling family was living in London at "Penlees," 40, West End Lane, West Hampstead: Ernest and Florence, with Muriel (age 21) Phyllis (20), John (16), and Ursula (14).

Ernest's response to the outbreak of war was an overwhelming feeling of treachery, of having been let down by the Germans; he had after all, always been a great admirer of everything German. He vowed he would never speak the language again, and began taking French lessons. (Throughout the war he had several French teachers, but none seemed very impressed with his accent.) He wanted to volunteer as a private soldier, but his friends pointed out that a 48-year-old professor was unlikely to be accepted. A month after the beginning of the war he wrote to August Krogh in Denmark:

> As there is nothing for me to do in connection with the war, I
> have to resign myself to going on with my physiological work . . .
> Of course the economic pressure in Germany has hardly begun
> to tell yet. It will however begin to tell by next summer when we

105

shall have an army of a million ready to put in the field. It would interest you to see the material of our new army. It represents the pick of our youngsters between 19 and 35, all as keen as mustard and learning the art of war as a sport. My boy is desperately keen on joining the army, but he is not yet old enough. I have promised him that he shall join next summer . . . But I expect I shall settle down in a little time to the novel mental conditions imposed on a non-combatant. Over two-thirds of our men (at University College) have joined the army in one form or another, either as privates or officers. (Starling, 1914)

There is a painful scoutmasterly enthusiasm here: "all as keen as mustard and learning the art of war as a sport." Would he think of trench warfare as "sport"? "My boy is desperately keen to join the army, but is not yet old enough" (he was 16). This was a time when it was believed that the war would be over in a few months, and the nightmare of trench warfare had not yet begun.

Starling applied for a commission in the Royal Army Medical Corps, and was accepted. He decided to work at a military hospital as a physician, and returned to Guy's hospital for a few weeks to revise his clinical skills. He did this with his old chief, (Sir) William Hale-White, whose house-physician he had been in 1889. Hale-White wrote: "When war broke out, he came, saying that he proposed to go round the wards with me to furbish up his medicine. Although he had done no clinical work for twenty-four years, he picked it up at once; in a fortnight he was as good at it as when he finished his house-physiciancy. He had a marvellous flair, not only for the science, but also for the art of clinical medicine" (Hale-White, 1927).

So Captain Starling became Medical Officer at the Herbert Hospital, at Woolwich in East London. But his clinical career lasted only a few weeks; early in 1915 he was asked by the War Office to run a research unit investigating poison gas. The unit was at the RAMC, Millbank, and he was promoted to Major.

Angry Gas Man at Millbank

The Germans first used poison gas—chlorine—at Ypres, on April 22, 1915. The British and French were unprepared; a German deserter had actually warned the Allies of the imminent attack, but he had not been believed (Haber, 1986). Two thousand prisoners and 51 guns were captured by the Germans in the operation. It took the British three and a half months to retaliate—they released a chlorine cloud at Loos, in August 1915. Starling's research unit (along with several other laboratories in the country) was investigating some of the candidate poison gases, but the overall British strategy could only be called chaotic. There were committees set up by the cabinet,

Figure 5-1. Major Starling, about 1916. (*Family Collection*)

the War Office, the Royal Society and the RAMC, each with a different agenda. What was the relative importance of defense and offense in gas warfare? The candidate gases included chlorine, phosgene, chloropicrin, diphosgene, hydrocyanic acid, and mustard gas. There was little agreement about whether gases should be sprayed as a cloud from cylinders (in the way that the Germans had used chlorine) or packed into bombs, shells, or cartridges and shot in the direction of the enemy.

May 1915 also saw the Germans torpedo the Cunard liner *Lusitania,* drowning over a thousand civilian passengers. Krogh and Starling exchanged outraged letters: "a shocking revelation" (Krogh, 1915) and "no victory will be of any use that does not mean absolute annihilation of the Prussian power—and that implies going on to the bitter end" (Starling, 1915a). The Oxford physiologist J.S. Haldane (brother of Lord Haldane, of the Royal Commission), who was working on poison gases in Oxford, published an official report on the use of chlorine by the enemy. He said the use of gas and the sinking of the *Lusitania* had finally convinced the public of the atrocity of the Germans.

In October 1915, Starling ended his research into what might be called the offensive use of poison gases, and was transferred into defense. He was still at Millbank, and his group had the coded name of "Hygiene," a curious expression that was borrowed from the Germans. Basically, Hygiene involved teaching the British army how to defend itself against gas. More specifically— how to use masks, which were known as "smoke helmets" (Haber, 1986).

The first (anti-chlorine) masks used by the army had been pads of cotton-waste soaked in sodium thiosulphate ("hypo") solution—they were tied around the soldiers' heads with strings. It was exhausting breathing through them. The first improvement to these was a complete permeable cover for the head, with two round celluloid windows for the eyes and an expiratory valve for the mouth. The fabric of these helmets was flannelette—pajama material—soaked in hypo before use, which was reasonably effective against

Figure 5-2. Stages in putting on "gas helmets." The unfortunate users breathed in through the flannelette helmets, which were soaked in a variety of solutions to counteract poison gas. Expiration was through a valve in the face of the mask. The type shown is the PH (phenate-hexamine) helmet, which was soaked in a solution containing phenol, caustic soda, glycerine, and hexamethylene-tetramine. (*Wellcome Library, London, with permission*)

chlorine. For use against other gases, such as phosgene or tear gas, the helmets were soaked in hexamine (a Russian invention). In the absence of hypo or hexamine, the men used urine, whose efficacy was unknown.

Starling's first letter to a superior officer, Colonel William Horrocks, Head of Gas Strategy—is positively upbeat. He was to write nothing like this again:

> I have been lecturing to young officers at Chatham, Dover,
> Colchester and Felixstowe . . . the present helmet affords perfect
> protection against practically all the gases which the Germans
> could use, but they are of no value unless the men know how to
> use them, and especially how to breathe . . . some of the officers I
> have lectured to would be quite capable of instructing if they
> could be provided with practice helmets for the purpose. I would
> like to recommend:
> 1. Six lecturers appointed to give lectures and demos.
> 2. 5–10% of men lectured should be given practice helmets
> 3. Medical Officers should be trained in the use of helmets; I
> am prepared to give a lecture/dem. one afternoon a week at
> Millbank, say 50 MO's per day (Starling, 1915b)

He got his way with all three suggestions, and he appointed four physiologists as the first lecturers. They were all established researchers: Francis Bainbridge, Edward Cathcart, Bertram Collingwood, and Charles Lovatt Evans. All four men were, or became, professors of physiology, while Bainbridge, Cathcart and Evans became Fellows of the Royal Society.

But not much went right with Starling's plans. His gas teachers, who were very enthusiastic at first, often had to lecture to up to a thousand troops at a time, without the use of microphones; we have no idea of why this should be. There was a permanent shortage of practice helmets. When these did arrive for teaching sessions, the valves were often faulty. Thus Starling, writing to Colonel Horrocks (February 22, 1916) complains: "while lecturing at St Albans there were 39 practice helmets. The valves were hopeless and should never have been issued . . . they were not up to sample—they ought to have been rejected at the first, and the whole lot dumped on the man who passed them—yours etc. (Starling, 1916a).

He complains to a higher superior (General Atkins) on May 26, 1916: "Is there no-one responsible for gas training who can stir up matters as regard supplies? Why cannot the army council put somebody definitely in charge of this whole gas business? At present no-one seems responsible for anything" (Starling, 1916b). Since Atkins was in charge, and Starling knew this well, these remarks must have been thought magnificently insubordinate. Atkins's reply has not survived.

Other, less likely, problems in gas education are aired in this letter from Lovatt Evans to Starling. Although Evans had worked in Starling's laboratory for three years, he properly addresses him as "Major." What Evans says, though, could hardly have been said to a normal superior officer (Evans, 1916):

Edinburgh 20/5/16

Dear Major Starling,

All CO's and officers are decent to me . . . NCO's are splendid lads and awfully attentive. The worst disciplined of all are, of course, the RAMC men . . . often more than 500 at lectures . . . I try to keep it simple so as to reach the men rather than the officers.

Your wire to my last port of call made a wonderful difference to my reception. Previous to that I might have been a lunatic wandering at large in his pyjamas for all they seemed to care or know. They have a new staff major here whose chief aim in life is to make a monocle stick in its place, and the best man for Collingwood to go to see is the Sergeant Major in the general staff office. [Collingwood was about to replace Evans as gas lecturer in Scotland]. Would you tell Collingwood? Of course he must report to the Colonel, who doesn't care a cuss about gas, and only wants to get you out of the way, or to the monocle, who neither knows nor cares. Until yesterday it has rained almost continuously.

Yours sincerely,

C. Lovatt Evans

The correspondence of the Hygiene department makes depressing reading. Without stinting his language, Starling continues to complain about most aspects of gas policy. In May 1916, he even has to complain that his gas officers are being claimed back by their medical schools, with the army making no attempt to prevent this:

Barts are trying to get Bainbridge back from the army. We need men like him—responsible for training about 100,000 troops in anti-gas measures. This training is more important at the present time than the preparation of a score of students—mostly crocks— for the Conjoint Examination [a medical qualifying exam, now extinct], especially as his assistant at Barts is doing no war work and can give all his time to the men.

Perhaps I take a biased view of these things. UC allows me to give my whole time to this job, and still pay the greater part of my income . . . Barts does not live up to this standard. London hospital will follow suit and I shall lose Cathcart as well as Bainbridge. It is not easy to find men with the right qualifications . . . (Starling, 1916c)

Bainbridge had been appointed Professor of Physiology at Barts in 1915, and spent the rest of the war between Barts and the RAMC; Cathcart's wartime career was similar. Their loss from his group of lecturers increased Starling's disillusionment. In the summer of 1916, according to Lovatt Evans, "Starling was proposing the use of mustard gas as an offensive weapon; this was not part of the British plans, and his suggestion was rejected. Then Starling made a vigorous protest at the highest level." Some months later, in July 1917, the Germans began using mustard gas shells.

Starling's protest sounds as though it was the last straw for the War Office—they had been hearing interminable complaints from this man for some time. In the inscrutable manner of the military, he was promoted to Lieutenant Colonel, and sent to Greece. Salonika, in northern Greece, was the center of the Balkan Front, a relatively insignificant war zone, and temporary home to about 90,000 British troops. Starling was to provide them with gas education. Since most fighting in the Balkans was in mountainous country, gas was unlikely to be used, and there is actually only one occasion on this front when its use was recorded. It seems as though Starling had been given a non-job, a sinecure, to keep him quiet.

More Gas: in Salonika

The front in Macedonia meandered, at some 30 km distance, around the north of the city of Salonika; the front was opened in November, 1915 (Palmer, 1965). It represented the line of confrontation of the Allied Entente powers (Britain, France, Serbia, Italy, Russia, and, sometimes, Greece) in Salonika against the Central powers (Germany, Austro-Hungary, Turkey, and Bulgaria) to the north. The British and French troops arrived in Salonika in October 1915; there were 90,000 British and 60,000 French, and it was the British faction that Starling was to educate in the ways of gas, beginning in November.

The Anglo-French force dug trenches and put up barbed-wire entanglements for the Macedonian front, transforming Salonika into one of the most fortified cities in the world. The chief fear was attack on the northeastern sector by Bulgaria. There was certainly desultory shelling by the Bulgars, which was postponed when the British were playing football matches. The expected big attack never came. At this time, soldiers thought that being posted to Salonika was a soft option, especially when compared with the Western Front; even music-hall songs of the time reflected this.

On November 9, 1916, Starling passed through the Mediterranean on the troopship *Canada,* on its way to Salonika, zig-zagging to avoid the attentions of U-boats (Starling, 1916d). We have a series of letters to his mother at this time; their tone is remarkable. Starling could have been appalled by his treatment from the War Office. But his letters show no hint of pique— all he says as he arrives is that he would like to get a gas school going as soon as possible. Before the *Canada* had reached Salonika, he had cabled off requests for vital supplies that he is sure will not be available in Greece.

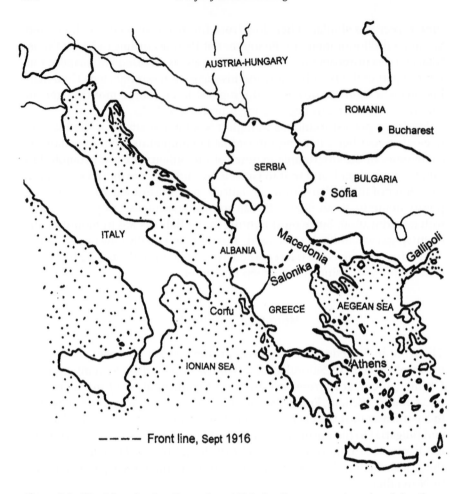

Figure 5-3. The Macedonian Front, late 1916. Starling was stationed at Salonika, and was responsible for the gas education of about 90,000 British troops. Gas was rarely used against them in this campaign.

Typewriters loom large. His letters all have the censor's stamp on the envelope, but there is no evidence of a censorial hand in the letters themselves. Presumably the ordering of typewriters from the other side of Europe was not information thought likely to be useful to an enemy.

Inevitably there is some anger in his letters, but it is directed towards the government rather than the Army. He tends to find the government responsible for German atrocities: "Much excitement here at the torpedoing of hospital ships. They are delightful people these huns. I wish we could clear out Parliament and Government and start afresh, but probably all the men we want in Parliament are serving somewhere as Lieutenants in the Army or Navy" (Starling, 1916e). These remarks were prescient, for a fortnight later, Lloyd George replaced Asquith as Prime Minister, which certainly had Starling's approval. This was at a time when, on the Somme, on the Western Front, 420,000 British troops had died in 4 months.

He was billeted in a picturesque house on a peninsula, jutting out into Salonika harbor. In describing the house he reveals some new sybaritic aspects of his personality:

> But we have electric light, and I have made my servant fill my water bottle with hot water and put it into my bed—so I can get my toes warm . . . I make a little pot of tea on a Primus stove every evening and drink it in bed. It is now beside me. I have bagged a brilliant scarlet rug for the side of my bed—so my room is more gorgeous than ever (I am wearing blue silk pyjamas—beloved of Muriel—but quite in keeping with the room) . . . I ought to be looking out onto the moonlit sea, as my bedroom has windows in three of its walls, but it is a cloudy night, so the moon will not show herself and all one sees is the solitary light on each ship— occasionally a warship twinkles as it sends messages to the signals officer on the white tower." (Starling, 1917a)

This Turkish tower, whitewashed and about 120 feet high, affords him pleasure in the early morning: "The sun rises very late now—about 8—so as I wake and look out of my right hand window—all I see is a bluish mist— and then a magic town all in pink appears out of the blue layer as the white tower catches the first red light of the rising sun. But then it gets light very quickly, just as it does in the east" (Starling, 1917a).

Starling writes of shopping in Salonika, after hearing that antique Turkish brass is a good buy. He doesn't find any, but discovers that the local Turkish delight is "the best ever tasted," and sends a box off to his daughters. (In a letter his mother confesses that she opened the parcel and scoffed the lot.)

For these few months, his life seemed idyllic." . . . I sometimes think it must be quite wrong that I like my job here so much I took a walk on to the hills this morning—I smoked a cigarette lying in a patch of grass with a glen, and a stream with croaking frogs and mosquito larvae hatching out in the pools in thousands. So far the mosquitoes are not troublesome— have sent my mosquito net to the wash. It is a beautiful rose-coloured one . . . but I shall have to put it up as soon as the weather gets warmer"' (Starling, 1917b). (Not many men on active service can have been so concerned with the cleanliness of their mosquito net.) Malaria was a serious problem on the Macedonian Front; in 1917, a quarter of the British force was in hospital with the disease.

His enthusiasm for the French language had not waned. He discovered a teacher of French called Mrs. Paschalis—an English lady who was the daughter of a French doctor in Cannes, "who was fool enough to marry a Greek, but has now got rid of him. Just imagine marrying a Greek . . . Of course the war is a boon to her . . . she gives lessons to quite a number of the GHQ staff—some of whom take a lesson simply for the pleasure of her company." Starling arranges for a map-making friend, Major Wood, to pick up Mrs. Paschalis and himself in his Ford van; Major Wood then drives them to the high hills behind Salonika. "Then we get out and walk home, first on the level and then downhill, getting a fine sunset en route—though for the last few days there has been a heat haze in the evening and not much view. But the walk goes along the whole ridge of hills behind the town, and finally descends onto the citadel . . . there is little snow on Mount Olympus, so the views are not so good as a month ago. Before church I spent an hour ironing my clothes" (Starling, 1917c). Starling seems to be simultaneously engaging his enthusiasm for mountains, for speaking French, and—who can tell?—for Mrs. Paschalis.

He decided, in June, that he had finished his work in Salonika. No one seemed to disagree. He had, presumably been lecturing on gas as he had been at Millbank, though in none of his letters does he mention any details. The troops no longer had to protect themselves with sodden flannelette helmets, for the design of gas masks had moved on. The next model was known as the small box respirator (SBR) introduced at this time, and involved a comfortable, impermeable, face covering, with a rubber tube passing to a perforated box. This contained a filter, of which charcoal was the main component. It was a design that hardly changed for the remainder of the Great War and for

Figure 5-4. Stages in putting on the Small Box Respirator (SBR), a great advance on the PH helmet. Its face covering was impervious, and inspiration was via a flexible tube ending in a perforated container containing charcoal. The container stayed in its canvas bag during use, and the user breathed out via a valve. (*Wellcome Library, London, with permission*)

World War II. The mask, tube, and filter-box were kept in the flat, square canvas bags that proved ideal for carrying sandwiches.

Starling returned to England, via Italy, in early July 1917. The night before he left, he dined with General Milne (the Commander-in-Chief). "He seemed very satisfied with my work—agreed that it was finished . . . "(Starling, 1917a).

Milne had a deteriorating relationship with his charismatic French counterpart—General Maurice Sarrail. Sarrail was actually head of the Anglo-French force in Salonika, and had commanded several unsuccessful attacks on Bulgaria in 1916–18. Because Sarrail had supervised his forces in so much trench-digging, the French politician Clemenceau famously christened the Anglo-French force as "The Gardeners of Salonika." Starling met Sarrail once, in December 1916: "A tall man, grizzled, with restless eyes. They say he's a better politician than soldier—but I don't know enough to judge" (Starling, 1917a). Clemenceau finally removed Sarrail from his post in the summer of 1918, and it so happened that the final successful push into Bulgaria occurred in September—it had certainly been begun by Sarrail. Within a few weeks, Bulgaria had fallen, and Germany's soft underbelly was exposed; early in October, Count von Hindenberg asked Woodrow Wilson of the United States to oversee an armistice. The Great War had begun and ended in the Balkans.

Meanwhile, Starling was driven in his army Ford car to six sites in distant corners of Italy. He gives his mother no clue of what he was doing. He was, in fact, acting as an agent for the British government, assessing Italy's capabilities in gas warfare.

It is not clear from Starling's report whether he was acting in response to recent events, for he writes that on June 29, 1916 there had been a gas attack by Austria on Italy on the Eastern Front. There were 7,000 Italian casualties, of whom 4,000 died. This terrible statistic suggests that the Italians had no gas masks; Starling's report confirms their shortage. He found widespread ignorance of gas defence among the Italian forces, yet he was complimentary about the country's technology in the making of poison gas. The design of Italian gas shells so impressed him that he advised the War Office to consider adopting them (Starling, 1917e).

His report must have had some influence on the War Office, for over the next year 300,000 small box respirators were sent from England. And by the middle of January 1918, arrangements had been made to supply 1.6 million more at the rate of 300,000 a week! (The Italians were presumably expecting the worst from Austrian attacks.) Starling was closely involved with these million gas masks, as we shall shortly see.

The Country's Food—The Royal Society Food Committee

Starling returned to London, via Paris, in July 1917. His department at UCL was doing research on poison gases, and managing to teach physiology to some 60 medical students a year (about a half of the number that would be taught after the war). Bayliss was acting head of the department, and the college was doing its wartime duty by providing public lectures. Bayliss himself gave several on "Pavlov" (which were very well attended) and "Food and the War Economy" (which were not). Marie Stopes, lecturer in paleobotany, gave several lectures on "Coal" (Annual Reports, UCL). Starling greatly admired Stopes's more famous persona (as an advocate of birth control and the sexual emancipation of women) and the admiration seemed to be mutual. Stopes wrote her most famous and influential book *Married Love*, during the darkest days of the war. In the words of a biographer, Keith Briant: " . . . as the news from the grappling armies on the Western front grew grimmer and grimmer, Marie Stopes took out the manuscript of her book again and decided to send it to Professor Starling . . . she felt confidence in it, but she feared when she sent it that perhaps he might find it unworthy . . . she longed for him to reply saying that it was good . . ." (Briant, 1962). Starling actually thought it outstanding, and his enthusiastic response became a preface to *Married Love*. We can hardly envisage now the extraordinary runaway success of the book, selling hundreds of thousands of copies; Starling's little preface surely obtained the highest readership of anything he ever wrote. Following the book's success, Marie Stopes gave up paleobotany at UCL, and Starling wrote to her saying what a great loss her departure was to the college.

Bayliss's wartime lectures aired a serious national problem: nutrition. German submarines were sinking food ships—many of them American—at a frightening rate. By October 1916, nearly two million tons of shipping had been sent to the bottom, raising a real fear of mass starvation. The Ministry of Food was established to bring some order to the feeding of the country; but its first head, Lord Devenport, was a disaster.

> The new Ministry of Food did not start well, partly on account of
> Lord Devenport's reluctance to accept scientific advice. He
> seemed unwilling to believe that a scientist could know more about
> food than one who, like himself, had spent a lifetime as a provision
> merchant, and he made many mistakes that aroused criticism both
> in official circles and in the press. No amount of hard work, for he
> killed himself overworking at his task, could make up for the
> blunders caused by his ignorance of the simplest facts about the
> functions of food. (Drummond and Wilbraham, 1937a)

Fortunately, some enterprising person in the cabinet proposed a countermeasure to the Ministry: an advisory group of scientists, physicians, and agriculturists. They made up the Royal Society Food (War) Committee (RSF(W)C), and were scientific heavyweights. They included Gowland Hopkins, Walter Fletcher (physiologist and later secretary of the Medical Research Council), Arthur Cushny (see chapter 7), Augustus Waller, and a

Glaswegian nutritionist, Noel Paton. Nutrition was a very new science—it had budded off from physiology, along with biochemistry, only very recently. Paton was important to the committee, having first-hand nutritional experience of the Glasgow poor. (For some reason, he communicated with the committee's distinguished secretary, W. B. Hardy, almost entirely by telegram.)

Starling attended his first meeting of the committee in November 1917. He is referred to in the minutes as "Colonel" Starling, though he actually resigned his RAMC commission at about this time. He had been to only two of the weekly meetings when he was made chairman in December 1917.

The committee spent a good deal of time correcting the mis-statements made by the Ministry of Food (Minutes, RSF(W)C, 1917). The Ministry had produced a "Food Economy Handbook" in which the public were instructed to chew their food three times, "for this enables more nourishment to be got from food"; "Before the war, the nation could live on the food it threw away" and "If you are eating meat you are better without bread: starch and meat together double the stomach's work." These physiological howlers aroused the committee's scorn, though this was always expressed most diplomatically by Hardy. Here is Hardy writing to Lord Devenport's deputy: "Both Lord Devenport and yourself are immersed in a medley of conflicting interest and you have to work in an atmosphere of interview and hurry which must make it extremely difficult to find time for critical consideration of possible causes of action in the light of fundamental principles" (Minutes, RSF(W)C, undated). Hardy's message is clear enough: you are out of your depth.

Under Starling's chairmanship, the committee increased its output of memoranda, and clear statements appeared on how food-rationing—assumed to be inevitable—should be organized. One of the committee's themes was that bread and cereals should never be rationed; rationing should be reserved for meat, fat, and sugar. In this way, however much daily work a man performed, he should be able to avoid hunger by eating cereals or bread. (One of the Ministry's unfortunate early memoranda stated that bread should be rationed *first.*)

But nothing is quite what it seems. Starling, in spite of bringing new life into the Royal Society Committee, didn't meet with Hardy's approval:

Jan 9, 1918
Dear Starling
    I feel that if we are to maintain our reputation as a business-like body when outsiders come, such as Yale and Gonner, you will have to take much more vigorous control of the discussions and steer them along. Kempe [Starling's predecessor as chairman] was extraordinarily good at that . . . we do not proceed enough by formal motion. We have lapsed rather to the stage of a group of friends talking things over by the fireside . . .
    Yours sincerely, W. B. H.
                                    (Minutes, RSF(W)C, 1918)

To put such thoughts in writing (could he not have just mentioned the matter to Starling?) seems the action of a very tough nut. Did he *have* to tell Starling that his predecessor was a better chairman than him? But whatever Starling may have lacked as a chairman, he clearly impressed the Ministry. For they saw in him a way out of the embarrassing relationship between themselves and the Royal Society committee: they would invite Starling to attend the Ministry's meetings. One month after Hardy's criticisms, Starling received the invitation; he was now chairman of the committee *and* Adviser to the Ministry. The Ministry were protecting themselves by having him in the room whenever they made a decision. It made Starling, for the last eight months of the war, the most influential person in the country's nutrition.

Food rationing, along with price controls, began in February 1918. The scheme rationed meat by value (it was equivalent to 1–2 pounds per person per week) and butter and margarine by weight (4 ounces per week) both by means of coupons. The word "coupon" (in the context of rationing) seems to have been introduced into the language by this committee. Moreover, rationing was remarkably successful.

> Perhaps the most striking feature of this time of rationing in
> England was the negligible extent to which evasion and illicit
> trading occurred, an experience repeated in the Second World
> War . . . rich and poor, the great hotel and the small east end
> eating house were treated alike, and it was no easier to get a
> little extra sugar in Mayfair than it was in Bethnal Green.
> (Drummond and Wilbraham, 1937b)

The committee went to some lengths to discover the food situation in Germany and Austria. There was virtually no food coming into these countries from outside, and rationing was far more draconian than in Britain. The daily allowance was several hundred calories less than this country— an actual difference of 28%. Consequently, malnutrition was rife: tuberculosis and dysentery were common, with dysentery having a mortality of 35%. Most relevant to the German war effort was a fall of 30–50% in the output of munition workers (Drummond and Wilbraham, 1939).

In 1920, Starling was invited to give a talk to the Royal Statistical Society on the food supply of Germany during the war (Starling, 1920). His paper restated many of the facts mentioned above, although he also leveled blame at German farmers, who hoarded a lot of their produce and established a black market, adding to the suffering of poorer citizens. But perhaps the most interesting point to emerge from the paper appeared in the discussion. T. J. Hirst (we know no more than his name) suggested that the malnutrition of Germany resulted from a few German ration books being brought to England, and several million forgeries of these printed before being returned to Germany. This, Hirst claimed, threw the whole system into chaos, because there was no way that all the forged coupons could be redeemed for food. Unfortunately, Starling doesn't comment on this brilliant scheme in the discussion. Could it have been MI5 or MI6? Was T. J. Hirst a secret agent? We will never know.

After the war it was discovered by British doctors that thousands of children in Austria and Germany were suffering from rickets or scurvy. Rickets was believed in Germany to be an infectious disease. The importance of accessory food factors (shortly to be called "vitamins") having been clearly demonstrated in England by Gowland Hopkins, Edward Mellanby, and Harriet Chick. The importance of vitamins had influenced the conclusions of the Royal Society committee, but they did not seem to have been accepted or acknowledged by Germany. This sad story signalled a fall from grace for German medical science, which twenty years before had led the world. It seems that the dire nutritional state of its people in 1918 contributed significantly to Germany signing the armistice.

Gas Masks for Italy—An Unexplained Episode

The complexity of Starling's life at this time is illustrated by an episode early in 1918. Food rationing was just getting under way when he was asked by the government to return to Italy on another gas mission, presumably arising from the contacts he had made there in 1917. The source for this story is Charles Lovatt Evans, who told it in the first Bayliss-Starling Memorial Lecture in 1963 (Evans, 1964). There are parts of Evans' account that are difficult to reconcile with official sources in the Public Records Office. His version certainly makes more interesting reading—Evans describes how "early in 1918":

> Starling was asked to go to Italy to convince the Italians by
> demonstration that ours was the best available [respirator] and to
> negotiate transfer of the requisite number of these. He asked me
> to be detailed to go with him, as I had seen gas in action, and had
> run gas schools . . . so we started out on the journey, by train,
> Starling a civilian, with a fur-lined coat, which looked and proved
> to be, expensive, our baggage being routinely rifled on route,
> and finally got to Rome, Padua and *Commando Supremo*. We knew
> no Italian, and the trip was in large measure a mixture of comedy
> and farce, but before we left, 1¼ million respirators had been
> delivered, and fitting begun.

It is all rather mysterious. Does "negotiate transfer" mean that the Italian government was buying the respirators? Where were the respirators when Evans and Starling were travelling? Would the respirators have been sent back to London if the negotiations had failed? The largest manufacturer of small box respirators (for that is what they were) was John Bell, Hills, and Lucas, of Tower Bridge Road, London SE1 (Document in the Wellcome Library); they began the manufacture of several million SBRs in July 1916 and were still making them in large numbers at the end of the war in November 1918.

Starling's report on the Italian gas situation had been written some seven months before. In it he had described the gas attack by the Austrians (in June 1916) that had led to 4,000 deaths in the Italian army. The official British report in the Public Record Office (by a Major Mackilly) says that by the middle of January 1918 "arrangements had been made to supply 1.6 million SBR's from England" (Mackilly, 1918). Could Starling and Evans' trip have been making these "arrangements"? It is not clear whether the Italians ever made use of this enormous collection of superior respirators.

Evans noticed that while they were having their negotiations in Rome, Starling would slip away in the afternoon, never explaining where he was going. When they left Italy, he was carrying a square box, about which he was equally uncommunicative. It wasn't until after Starling's death, nine years later, that Evans discovered that Starling had actually been carrying a bust of himself. Giulo Fano, Professor of Physiology in Rome, had persuaded Starling to sit for the bust, and that is where he had been on those missing afternoons (Evans, 1964). The bust (which Starling thought a poor likeness) is still in the Professor's office in UCL.

The melodrama of the visit continues on their journey home. Evans writes:

> Finally we left Padua, in an air raid warning, and were hurriedly
> bundled into the train; Starling then found that his passport was
> missing, and we were told at Milan that we must go back, but he
> went on, and at Modane I smuggled him across the frontier.
> After we reached Paris his passport caught up with him; this
> eased our minds until we tried, some days later to board the
> 7 a.m. leave boat at Boulogne, when it was spotted that his
> passport had no permit to leave the war zone. Thereafter I saw
> our luggage put on each successive leave boat, and taken off
> again. Eventually, after various formalities, we got the last leave
> boat of the day. (Evans, 1964)

During the last months of the war, it became clear that food shortage was not going to disappear with the end of hostilities, and an Inter-Allied Food Commission was set up. Starling was one of two British delegates, along with two from France, Italy, and the United States. An overall food policy was agreed upon—Starling summed it all up in an excellent little book (perhaps having a rather transient influence) called *The Feeding of Nations* (Starling, 1919).

In retrospect, Starling had a remarkable time in the war, providing the nation with two quite different scientific skills. The only reward for his services was the CMG (The Order of the Cross of St. Michael and St. George); this quirky decoration was said to be for his contributions in Thessalonika, where he actually did very little. Was the decoration some sort of governmental guilt for having sent him to Greece in the first place? His achievements with gas warfare and nutrition in London seemed far more worthy of recognition, but they brought him nothing. We shall see shortly how his wartime experiences seriously affected his attitude toward the government, and, in consequence, the government's attitude toward him.

6

1918–1920

The Old Anger: Classics and Class

During the last few months of the war, when Starling was spending most of his time back at UCL, the *Lancet* asked him to review a book. It was *The Report of the Committee on the Position of Natural Science in the Educational System of Great Britain*: hardly a riveting title. But it got Starling going—producing an explosion of all his accumulated fury at the government and the educational system. The review begins:

> The astounding and disastrous ignorance of the most elementary
> scientific facts displayed by members of the government in the
> war raised doubts in the minds of the British public as to the
> efficacy of the education imparted at a high price to members of
> the upper classes, including those who are chiefly responsible for
> the control of the destiny of this country. (Starling, 1918)

The elements of science are a closed book to those in positions of responsibility, he wrote, for our educational system is built round a knowledge of Greek and Latin. It is not that these subjects are unsuitable topics for study—it is just that so many years are passed at school and nothing is learned:

> After nine years, nine-tenths of the boys can read neither Latin nor
> Greek. They may have acquired a few catchwords or allusions to
> classical mythology, but they can give no account of the manner in
> which the Greeks lived, of the part played by Greek philosophy or
> art in the evolution of modern ideas, or of the way in which
> western government had been founded on Roman inventions . . .

Moreover, the pupils, whose studies have been almost entirely literary and grammatical, are incapable of writing with fluency in any language, including their own. The situation is worst in the public schools (in Britain, these schools are private) over which Starling is predictably scathing: "Men send their sons to public schools because there are advantages which more than outweigh the disadvantages of a senseless method of education." To go to a public school is to join a guild, a club, and being either a member of the ruling classes, or belonging to a guild, is much more important than any intellectual attainment. And here he reverts to his original scorn for the government: "It is, indeed, regarded almost as heresy to demand a government minister a special knowledge of the work he is appointed to direct, and the idea of promotion by merit in the army or other public service arouses feelings of horror in the majority of these services." Starling had firsthand knowledge of the workings of the War Office and the Ministry of Food.

His conclusions were that the educational system needed total restructuring. Children should all have a general education until age 16, and from 16 to 18 should concentrate on a group of subjects that depend on the child's tastes and their proposed career. (These conclusions were anticipating School Certificate and Higher School Certificate, which appeared 10 or 15 years later, and did a lot to remove the obsession with Latin and Greek.) Starling did not propose a cure for Britain's class system.

The establishment (especially those members of it responsible for the awarding of honors) must have been startled by this attack. On a previous occasion, Starling had launched into the Harley Street establishment—see for example, his beginning of term address at UCL in 1903 ("He may now become a respected West End consultant; he will never add anything to the science of medicine"). It becomes easy to see why, for apart from the strange CMG, he received no military or civil honors in his lifetime. Yet William Bayliss, whose scientific achievements were less than Starling's, and was too old to play an active part in the war, was knighted in 1922; Starling's colleagues and contemporaries (Thomas Lewis, Henry Dale, Edward Schäfer, Thomas Elliot, Charles Sherrington, and Charles Martin, for example) were all rewarded in this way. Starling's original attack on the Harley street establishment had been accompanied by strong pro-German sentiments, and although he abandoned these during the war, I suspect that the Harley Street establishment stored these away as a component of their disapproval.

There is a remarkable precedent for his attack on the British educational system. Edward Schäfer had addressed a meeting on "The Neglect of Science" at the Royal Society on May 18, 1916. It was an onslaught on classics in the educational system:

> and we ask ourselves—looking over the circle of our acquaintan-
> ces at those who have had the irresistible privilege of having
> Greek and Latin swished into them from the earliest years—
> whether in the great majority there is any sign that there was ever
> much penetration beyond the skin . . . The claim has recently
> been made that without knowledge [of the classics] we are unable
> to express our ideas in our own language. The absurdity of this
> contention is obvious. . . . (Schäfer, 1916)

Here Schäfer cites Shakespeare, Bunyan, John Bright and Thomas Huxley as four masters of the English language who were without the advantage of a classical upbringing. He touches on Starling's *bête noir*, politicians: ". . . even politicians should know something about the world they live in and the bodies they inhabit. Surprise has been expressed at the singular ignorance displayed by distinguished statesmen of simple facts in chemistry and physiology . . . but what chance have they had to acquire any knowledge of these subjects?"

Schäfer's *Nature* article had been published in 1916. Had Starling seen it? Schäfer's attack was not as outspoken as Starling's, for Schäfer actually sympathized with unfortunate politicians, who were laboring under the handicap of a classical education. But there was little forgiveness in Starling's withering onslaught. It seems likely that this outburst, more than anything else in his life, prevented him from receiving a knighthood (though his letter to General Atkins in 1916—"Why cannot the army council put somebody definitely in charge of this gas business?"—must have run it close). There is no suggestion that this deprivation ever bothered him.

Notes from the Family

By the end of the war the Starling brood was fleeing the nest. His son John had joined the army in 1915 and stayed on in 1918 as a cavalry officer; he spent several years in Ireland, engaged in horsey activities. Ernest, though very devoted to John, was not impressed with his way of life.

With the usual entanglement of research and family, Muriel Starling, the apple of Ernest's eye, married Sydney Patterson in October, 1919. Patterson, who had been a coworker on the heart–lung preparation, had spent the war as a pathologist (at Rouen) in the Australian army. Presumably on the basis of his Beit fellowship and his published work with Starling, Patterson was appointed the first Director of the Walter and Eliza Hall Institute in Melbourne. We imagine that Starling gave Patterson a good reference, but, by doing so, was signing his Muriel away. The wedding in

October was closely followed by their journey to Melbourne in November— at that time the long voyage meant that parting relatives might never see each other again. Florence Starling finishes her first letter to her departed daughter: "Well darling, I hope this is gossipy enough—I've not put any questions to you as it only fills up the page without giving you any information, but always realise I want to hear every little detail, woes as well as joys. I think of Pat with the greatest affection and sense of repose for you. I could not have borne your going to Australia with anyone but him. Goodbye now for the moment, darling—Mother."

The second Starling daughter, Phyllis, also married at about this time. Her husband was an Irishman, Maurice Trouton, and they moved to East London in 1919: 222 Eglinton Road, in Woolwich SE18—a large, hospitable house that provided a haven for needy Starlings over the years. The Troutons were unusual in having no scientific connections. Finally, the youngest daughter, Ursula ("Babs") remained at home; she was studying for a diploma in Dairy Bacteriology at Reading University, and remained unmarried until after Ernest's death in 1927.

With the departure of the children, 40 West End Lane became too big, and it was put on the market in August 1919, for £4,000. It stayed on the books of the agents Goldschmidt and Howland until February 1920; it was finally sold for £3,500 to a Mr. Espelby, who, to Florence's delight, was manager of Steinway's English piano factory (Florence was still a serious pianist). Meanwhile, Florence, Ernest, and Babs began renting 23 Taviton Street, near the back door of University College.

Ernest was kept very busy in his laboratory, though not with research. He was Pre-Clinical Dean, he was short of qualified staff, and consequently doing a great deal of teaching; he was very relieved when his old assistant Anrep appeared from Russia. Starling signed him up for the January term of 1920, and returned to preparing the third edition of his *Principles of Human Physiology*, which was probably the most successful English textbook of the time. He was greatly helped by Florence, who was proofreader and indexer; she proudly writes that Ernest accepted about two-thirds of the changes she made on the proofs. This edition actually had two sets of proofs, which, for a book of more than a thousand pages, represented hundreds of Florence-hours.

But Ernest was feeling very tired, and characteristically decided that he needed a walking holiday. Unfortunately, it was December, but he persuaded Charles Martin to go with him to Cornwall. Martin was head of the Lister Institute in London, and the Institute had a vaccine station on a farm at Hayle, in Cornwall; Visiting Hayle was Martin's excuse for a holiday. As Starling wrote to Muriel: "We shall go there on the 29th [December] and stay at the farm, run by the excellent Green. He promises us cream and butter and a cask of claret—washed up by the tide: old Cornish fare in which the work of the wreckers has always figured largely. This time it is the Bosch that did the wrecking." They walked in the region of Kynance and Mullion, in serious south-west winds and rain, and by the time they were back, Starling felt even worse.

The St. Thomas's Scheme: Rockefeller Generosity

Some of the most eventful months of his life followed. The Haldane Commission's recommendations, as we saw earlier, had been shelved because of the war; the recommendations had included the establishment of clinical units in London teaching hospitals. And as we saw in the Commission's report, Starling was a firm believer in the possibilities of clinical research; so although he was not primarily a clinician, it was felt that his grasp of clinical medicine, along with his research expertise, made him a strong candidate for one of these new positions. Three Professors of Medicine were to be appointed. The names proposed were Archibald Garrod (St. Bartholomew's), Starling (St. Thomas's), and Thomas Elliott (University College Hospital). Florence Starling writes to Muriel, on November 20, 1919 "Turney [St. Thomas's secretary] says that the treasurer and staff were unanimous in offering him the appointment—the next step would be an interview with the treasurer re. Accommodation and ways and means." But Ernest's letters suggest that he was being persuaded against his will, for he was very uncer-

THE UNIVERSITY UNIT AT ST. THOMAS'S.

WE understand that Dr. E. H. Starling, C.M.G., F.R.S., Jodrell Professor of Physiology, University College, London, has accepted an invitation to become the director of the medical element at St. Thomas's Hospital, and that arrangements are in progress for the appointment of assistant directors in pathology and clinical medicine to work with him in the new element. We announced some time ago that Sir Cuthbert Wallace, K.C.M.G., surgeon to the hospital, had become director of the surgical element, so that two-thirds of a university unit has now been formed at this great hospital and medical school. Dr. Starling has been concerned for the greater part of his life with physiology as an investigator and teacher, but he has always regarded it and taught it as one of the institutes of medicine. During the war he

Figure 6-1. The Job that Never Was. (An announcement in the *British Medical Journal,* Jan 3, 1920.) Starling was appointed one of the first three professors of medicine in London. He never took up the post, partly through having cold feet over teaching clinical medicine.

tain about teaching clinical medicine. A first assistant (Ellis) and a chief of biochemistry (Maclean) were named for the unit, and on January 3, 1920, an announcement was made that Starling had been appointed (*British Medical Journal*, 1920).

By a remarkable coincidence, on the day following the announcement, two important American visitors, representatives of the Rockefeller Foundation (Drs. Wickliffe Rose and Pearce) were visiting UCL. The Foundation, here an extraordinary fairy godmother, was preparing to offer the hospital and college about a million pounds—partly to provide extra beds and clinical laboratories for the proposed medical unit, and partly for improvements in the pre-clinical medical school. Starling (as Pre-Clinical Dean) had already written a report stating that the departments of pharmacology and anatomy were particularly in need of modernization, and it now seemed that this modernization would be included in the Rockefeller gifts. (In fact, the anatomy department benefited, but not pharmacology). They really *were* gifts—almost no conditions were attached to the money.

Starling was now very confused—he had accepted a job that seemed inferior to his (suddenly improved) old job. On January 8, 1920, Florence and Ernest invited Charles Martin to dinner, for advice; he was always full of sense. Here is Florence's description of some of the conversation of that evening: "But this time they [the Rockefeller] want to do something in education in England and they mean this should be UC in connection with Ernest . . . Physiol Institute all right as it is, but pharm. must be enlarged and developed, and the anat. place is "wretched" were remarks dropped by Ernest. (He stipulated £2000 for the professor's salaries—they jibbed somewhat at this, for they don't pay professors too handsomely in the US, in their *palatial* university buildings) but Ernest insisted on the point . . . There was a discussion of Ernest getting out of his liabilities with St. Thomas's. Charlie Martin dropped a pregnant remark: 'I think the Americans are providence, intervening to save you from an act of *tomfoolery.*'"

So Charlie Martin had his way; tomfoolery was avoided, and Starling withdrew from the St. Thomas's chair. In fact two of the first three medical professorships (units) in London got off to poor starts, because Archibald Garrod, appointed to St. Bartholomew's, stayed only for a few months; he went to Oxford to succeed William Osler, who died in early 1920. Only Elliott at UCH stayed, his life undoubtedly made more bearable by the Rockefeller generosity.

Abdominal Pain: A Trip to India

Starling had still not returned to research work, even though the war had ended over a year before. As well as tiredness, he was now having bouts of abdominal pain, and toward the end of January 1920, he went with Florence to consult his old chief (now Sir) William Hale-White, at Guys. Florence wrote:

> . . . because he thought the pain might be due to appendicitis.
> However, Hale-White reassured him on this point. I got into the
> room a few seconds before Ernest, and managed to whisper to H.
> White: "Ernest is coming to consult you: knock off the cigarettes."
> But apparently he didn't catch what I said, as he didn't mention
> the subject . . . but merely informed Ernest that it was not
> appendicitis. But I had also told Lady H-W about the smoking
> and the next morning, H-W sent a little note to the lab, advising
> Ernest to smoke less. Well, he smoked not at all for 2 days . . .

This story tells us more about its astute author than any of the characters. Did Florence *really* say "knock off the cigarettes"? Did Hale-White understand her? She obviously had more insight into the dangers of smoking than any of the medical men who surrounded her.

Ernest's pain didn't improve, and Florence refers to it in letters as his "floating kidney." It isn't clear who made this old-world diagnosis, but events were interrupted by a letter from the India Office to the Senate of the University of London. The letter stated that Professor Starling should be invited to proceed at once to India. The purpose of the visit was to advise the Indian government on the best place and means of starting an all-India Research Institute. He found the prospect exciting, even though the Rockefeller offer to UCL had not yet been completed. Believing that a sea journey would restore his health, he accepted the invitation, and on February 20, seen off by his son John, sailed from Tilbury on the P & O liner *Moria*, bound for Bombay.

Starling enjoyed the sea trip, though his pain remained. He gives a memorable description of the ship's fancy dress ball:

> A little Scotchwoman [sic]—Mrs Wilson—dressed me up as an
> Arab woman. Only my eyes are visible—the face from nose
> downwards being covered by a black veil (made out of a silk
> stocking) from the top of which a gold cylinder goes to the
> middle of the forehead. Round my neck and arms and ankles,
> bead necklaces of all colours, and silver bangles and a long blue
> cloak over all, coming lower on the forehead. No-one spotted me
> [!] and I was awarded second prize. I also danced three times—
> with girls not superior to waltzing . . . I had to talk falsetto all the
> evening and play the part of an Arab bride ravished by Chu Chin
> Chow . . . two breakfast rolls made an excellent bust . . . I am very
> disappointed not to have had a cable from Mother—I suppose it
> means there is nothing to send . . .

But there *was* something to send, because the Rockefeller Foundation had cabled George Blacker, the clinical dean of UCH:

NEW YORK VIA WESTERN UNION 2ND MARCH 1920
BLACKER UNIVERSITY COLLEGE LONDON
ROCKEFELLER FOUNDATION GREATLY INTERESTED IN
PROJECT STOP CERTAIN QUESTIONS OF POLICY AND
ESTIMATES MUST BE SETTLED STOP CONFERENCE IN THIS
COUNTRY ESSENTIAL STOP COULD STARLING ELLIOTT AND
OTHER PERSON FAMILIAR WITH COLLEGE AND UNIVERSITY
RELATIONS COME TO NEW YORK LATE IN APRIL OR MAY AS
GUESTS OF ROCKEFELLER FOUNDATION STOP . . . NO PUBLIC
ANNOUNCEMENT STOP INFORM STARLING
PEARCE

The Foundation didn't realise that Starling was actually somewhere in the Indian Ocean. Their problem lay in understanding the complex relationship between the university, the medical school, and the college, and, before finalizing their gift, the Rockefeller wanted to discover exactly who they were giving their money to (Merrington, 1976).

Meanwhile, Florence was enjoying herself as a politician. Who was to go to New York? "Both Carey Foster [the Provost of UCL] and Blacker interviewed me as to Ernest's probable wishes in the matter and I said he would wish the Provost and Elliott to go—Blacker told me he meant to go too—that is quite right, as he is Dean of the Medical Faculty. Both Foster and Blacker laid stress on the necessity for secrecy . . ."

In the event, Elliott, Foster, Blacker, and Elliott Smith (the Professor of Anatomy) sailed. Elliott Smith took with him the plans for a new anatomy department that Starling had asked Simpson, UCL's architect, to draw up. The final Rockefeller gifts were £400,000 for the building program and a further £435,000 for maintenance; the college was also given £375,000 "for the teaching of anatomy on a scientific basis." These were the largest benefactions that UCL had received up to that time, and, allowing for the value of money, perhaps the greatest benefactions the college has ever received.

Ernest's tour of India was successful. His report to the government recommended building the all-Indian Institute in Delhi; it was subsequently built, and is there still. Unfortunately no copy of his report survives. He made a *faux pas* by forgetting to take his medal, which he should have worn at dinner every night. His abdominal pains persisted, and he also developed bouts of fever, which were diagnosed as malaria. He returned to England in early June, and went immediately to see Arbuthnot Lane, who was probably the best-known abdominal surgeon in the country. Ernest wrote to Muriel, saying that Lane was going to operate. And this was how his letter ended:

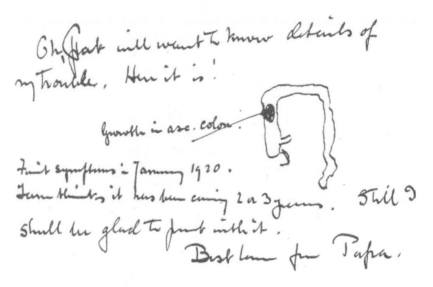

Figure 6-2. Starling's drawing of his colon

(Oh, Pat will want to know details of my trouble. Here it is. Growth in asc[ending] colon. First symptoms in January 1920. Lane thinks it has been coming 2 or 3 years. Still I shall be glad to part with it. Best love from Papa.)

The Operation

This is an extraordinary moment. Starling, with clinical detachment, is telling his daughter that he has cancer of his colon, and provides her with a precise sketch of his large bowel. If Lane were right, and the tumor had been present for two or three years, Starling's prognosis would have been very bad. However, this would depend on the malignancy of the cancer, and whether it had spread. We have no evidence on either of these.

Lane removed the tumor—he removed the right half of the colon (a hemicolectomy)—on June 15. The operation itself went smoothly, but recovery was greatly delayed by two episodes of pulmonary embolism. It is likely that this was the result of Ernest being kept in bed for too long after surgery; nowadays pulmonary embolism is largely avoided by getting the patient out of bed within a day or two of the operation.

The Physiological Society, at its July meeting, passed a vote of sympathy, and sent wishes for his prompt recovery. He recovered by going on holiday to Cornwall, and writes enthusiastically on the beauty of the Fal estuary: "I can't think why we didn't discover it before . . . here the crocks and the able bodies can find plenty to do—no crowd—little streams run-

ning everywhere—good walks and boating or sailing to your heart's desire. I came over the estuary (2–3 miles across) to St Mawes, when I have
hopes someday of having a family holiday again . . . don't you think you
had better come?" His deep feelings for Muriel are in every letter, often
accompanied by rather impractical invitations.

The operation marked a change in his holiday habits. Before it, he went
climbing—serious climbing with ropes and guides—usually once a year, in
the Alps. After the operation he never climbed again, but remained an

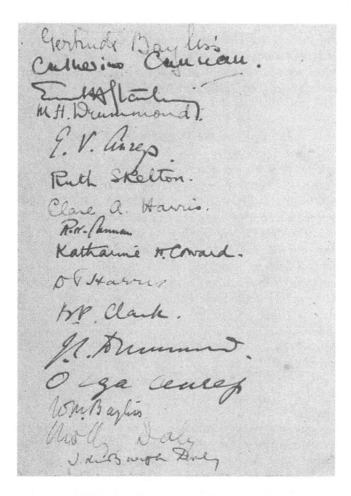

Figure 6-3. After Starling's serious illness in 1920, members of his department,
with their wives, welcomed him back with a dinner. The signatures are on the
menu from the Comedy Restaurant, Panton Street, London, in January 1921.
Some details of these individuals are included in Appendix II . (*From the papers of
Ivan de Burgh Daly, whose signature may be seen at the bottom.*)

enthusiastic walker. His basic plan was to go walking (often with Charles Martin) in a new part of the countryside, and look out for good-value Bed and Breakfasts. He would then try and center a family holiday around these, for the Starlings were never well off, and good cheap lodgings were essential. But he could hardly have had a less suitable family for his plans, for Florence and Ursula did not enjoy walking and John seemed always to be away with the army. Muriel enjoyed walking, but she was on the other side of the world. In October 1920, Muriel and Sydney produced the first Starling grandchild, Thomas John Starling Patterson, and, eighty years on, Tom Patterson has retired after a successful career in surgery.

Starling returned to work in January 1921, and on the 21st of that month the staff and their wives celebrated his return by entertaining him to dinner at the Comedy Restaurant in Panton Street. Burgh Daly writes of it as a great occasion, with Starling back to his best form. A signed menu from the evening has survived (Daly, 1967) (Figure 6-3).

The Birth of the Anatomy Department

The Rockefeller bequest was completed toward the end of 1920, and in December, the new Anatomy Institute (it was Starling's title, not the official one) was begun by demolishing a row of seven houses in Gower Street. Florence wrote, with her concern for the price of everything: "The seven houses in Gower Street adjoining the Physiological Institute are now all cleared away: only vast stacks of bricks remaining on the razed ground. The houses only cost £500 to demolish, because the sale of timber etc covered the remaining cost. On the other hand, cleaning and stacking of the bricks cost £1,250. New, they would have cost about £400. This bit of cockled paper is due to the tea, which spilt itself."

The T-shaped anatomy department, as proposed in the Simpson–Starling plans, adjoined the Physiology Institute. Its facade is neo-classical, and rather depressing, with a front door opening onto Gower Street. It lacks the grace and simplicity of Simpson's Physiological Institute, built fourteen years before. The anatomy building was officially opened by King George V in May 1923.

As well as opening the new anatomy building, the King and Queen also laid the foundation stones for a new obstetric hospital in UCH, along with a nurses' home (they were Rockefeller gifts). The king's speech naturally centered on the generosity of the Rockefeller Foundation, and expressed hope for the Unit system in London hospitals (Annual Report UCL, 1923–24). The elaborate unused doorway from the medical school library onto Gower Street was used—perhaps for the only time ever—for the Royal party passed through it, crossed Gower Street (with an academic "guard" lining their path) into the entrance of the Anatomy Department. There they had

Figure 6-4. The UCL anatomy building in Gower Street, on the occasion of its opening by King George V in 1923. It has changed little in the intervening years. (*College Collection Photographs, Library Services, University College London, with permission*)

tea in the Anatomy Museum with Elliot-Smith (the Professor of Anatomy), Simpson (the Architect), and Starling (whose idea the whole thing was). Very few anatomy departments can have been opened by royalty; we assume that the King and Queen were not shown round the dissecting room.

Lovatt Evans, Starling, and Dale: A Vignette

Charles Lovatt Evans was subsequently Starling's successor at UCL. He had a meteoric career in physiology, and this was in no small way a result of Starling's support. Evans became Sharpey Scholar at UCL in 1910, mainly by impressing Starling in a physiology viva, and by 1913 he had been awarded a DSc (the research degree that was an early version of the PhD) in physiology. In 1916 he qualified in medicine at UCH, at Starling's suggestion, and joined the RAMC. There he became one of Starling's gas officers, as we saw in Chapter 5. In 1917 he was appointed Professor of Experimental Physiology at Leeds, and in 1918 began work there.

But it was not a success. He found the salary and facilities poor, and he wrote to Henry Dale asking if there was a chance of working for the Medical Research Committee in London. Meanwhile, Starling had heard about Evans's intention of resigning from Leeds, and took a characteristically uncompromising view:

Figure 6-5. (Sir) Charles Lovatt Evans, taken at UCL in the 1920s. (*Family collection*)

Dear Evans

. . . . don't hand in your resignation. You have got to show that you can make good up to the limits and possibilities of the place you occupy [Starling then sums up his own difficulties at Guy's in the 1890's] I do not see that you are any worse off than I was, and you have got your training behind you . . . Every body of men who elects a professor looks to him not only to do good work but to alter his conditions so as to improve not only his department, but the whole institution with which he was connected. So, for God's sake carry on and make the most of it! I hope you will be able to come to town soon and have a talk with me.

Yours very truly, Ernest Starling (Starling, 1919)

Perhaps it was easy for Starling to give this counsel of perfection, for it was what he himself had done (twice). To a mountaineer, mountains are there to be climbed.

Evans then writes to Henry Dale, explaining his dilemma. In his fastidious prose, Dale replies:

> 140, Thurlow Park Road, Dulwich
>
> Dear Evans
>
> ... I am much troubled by Starling's protest, of which your letter gives the substance, but unless there is more behind it than I can see, I find no reason for supposing that he has any motive other than genuine concern for your future, and for that of

Figure 6-6. Sir Henry Dale (*Godfrey Argent Studio (Walter Bird) with permission*)

physiology in this country. I foresee that our MRC institute will
have to face the suspicion that it is depleting the field of the best
candidates for chairs, and having no regard for the interests of
education . . . I am anxious that our plan [getting Evans to work
with the MRC] should be done without arousing Starling's
resentment, even if we cannot get his unqualified approval. I
have as much reason as you to regard myself as under lasting
obligation to him. I have on occasion acted against his advice, but
he has always been generous enough to let it make no difference
to his attitude. He has never been wholly sympathetic to the
MRC, and I suspect that his own recent experience of the official
world has prejudiced him further [a totally accurate guess!] . . .
Don't imagine for one moment that I am wavering in the
determination that you shall come to us . . . I am only anxious
that you should lose nothing of the good will of Starling in doing
so. Write and tell me that I may go and talk to him. Also arrange,
if you can, to come to the Phys[iological] Soc[iety] meeting on
Saturday. We shall be delighted to put you up and you can see
Starling and talk with him.
  Yours very truly, H.H. Dale. (Dale, 1919a)

This is a glimpse into a long-gone, gentlemanly world. ("Write and tell
me that I may go and talk to him" is beautiful). Dale's concern and admiration for Starling is balanced by his determination to get Evans for the MRC,
and he succeeds in balancing the concerns. Two days after this letter, Dale
meets Starling by chance, and they naturally discuss Evans. Then Dale writes
again to Evans (Dale, 1919b), repeating his previous invitation to come and
stay in Dulwich and so talk to Starling.

You will find him very reasonable, but I doubt if you will shake
his attitude that this Leeds appointment is a sort of test of your
power to face difficulty . . . I am sure, however, that you will not
need to fear his resentment, should you fail to convince him
and act against his advice. He will probably advise you to use the
offer of the MRC post as a lever to improve things at Leeds. You
are quite entitled to do this if you wish, but I still hope you will
come to us in the end.
  Yours very truly, H. H. Dale

Evans did go to work with Dale at the Medical Research Committee (which
became the Medical Research Council in 1920) and he spent three happy years
there. He maintained a good relationship with Starling, asking him to act as
referee when he applied, successfully, to be the Professor of Physiology at
St. Bartholomew's Hospital Medical School in 1922. Four years later Evans
became Jodrell Professor at UCL. But we are getting ahead of ourselves.

7

Back to Research

The Heart on the Railings

Ernest, Florence, and Babs rented 23 Taviton Street—some 300 yards from the back door of UCL—and moved there in March 1920. The family may not have been well-off, but they kept a cook. Florence makes it sound a cozy family ménage, describing herself and Ernest as "Darby and Joan" with music around the piano in the evenings. ("Darby and Joan" is slang for a close couple; the expression originally described chains that shackled two prisoners together.) Ernest was feeling stronger, and as some of his energy returned, he began taking dancing lessons. He took Babs to *thé dansants*, which were all the rage. There were even occasions—charitable events, organized by Florence—when the Starling and Bayliss families all went to dances together (though it is not easy to imagine William Bayliss fox-trotting).

Ernest's dancing enthusiasm coincided with his return to research. Early in 1921 he was thinking again about the heart as a pump, and decided that he wanted to know what actually happened to a ventricle when it contracted. His plan was to photograph a working heart (in the heart–lung preparation) with an ultracinematograph. This was a device capable of photographing 300 frames per second, and was being developed by Nogues, a French scientist, and Starling had seen some of Nogues' pictures of birds flying. So, at

Figure 7-1. The Starlings lived here—23 Taviton Street, close to the back of
UCL—from 1920 to 1927. (*Photo by the author, 2003*)

Nogues' invitation, in March, Ernest went to stay in the Marey Institute in
Paris. As Starling arrived, Nogues' father died (he lived in the Pyrenees) and
Nogues disappeared, locking up his ultracinematograph. Starling was determined to get a moving picture of the heart, so he set up a heart–lung
preparation and tried to photograph it with a conventional camera and flood
lights. Unfortunately, these were not bright enough, and what happened
next is described by Henry Barcroft in his Bayliss–Starling Review Lecture
of 1976: "Starling had been adamant—the film had to be made. Taking his
scalpel he cut through the tissues connecting the heart and lungs to the
animal's body, and with the help of those present, carried the preparation
and its attachments into the forecourt of the Sorbonne and tied them to

the railings between the courtyard and the pavement. What I chiefly remember about the film was the look of amazement on the faces of the pedestrian passers-by when they saw the beating heart on the railings, only a yard or two away" (Barcroft, 1976). Barcroft had been a student when he saw Starling show the film—shot by an unknown French photographer—some 50 years before.

What was its point? The most reasonable explanation was that Starling was due to give a Friday evening talk to the Royal Institution in May, and that he needed a film of the heart–lung preparation for that. It was not appropriate to present an anesthetized dog to such an audience, so perhaps a film would be the next best thing—but a beating heart tied to some railings? He duly gave the talk ("The Law of the Heart") to the Royal Institution, but, unsurprisingly the *Proceedings* make no mention of a film (Starling, 1923–24). He never published anything from this curious episode.

Starling and the Kidney

Between the end of the war and his death in 1927, Starling spent the greater part of his research time investigating the kidney. The point of departure had been his important finding that hydrostatic and osmotic pressures oppose each other in the capillary. He had quickly appreciated the relevance of this relationship to the renal glomerulus, and in 1899—while still working at Guy's—had published a key paper (Starling, 1899). At this time there was uncertainty about whether the glomerulus separated the cellular and protein components from blood by simple physical filtration, or by some active process. The views were those of Ludwig and Heidenhain, respectively; they were also the views of these two on the formation of lymph (Chapter 2).

Starling's glomerulus paper assumes Ludwig to be correct; all Starling needed to do was to prove it. His initial experiment was to measure serum osmotic pressure—a slightly more sophisticated experiment than one that he had done while working on lymph. He began by separating protein from serum. This was achieved by filtering serum under a pressure of 30–40 atmospheres through a porous cell previously soaked in gelatine. The protein-free filtrate was then put into the inner compartment of an osmometer, with the protein-containing fluid in the outer compartment. This (after 2–3 days) gave rise to a pressure of 30–40 mm Hg, and represented a force of this magnitude acting from outside in the glomerular capillary. Starling then points out that 30–40 mm Hg of hydrostatic pressure is the lowest pressure at which urine flow is observed; below this pressure, the outward hydrostatic pressure becomes less than the inward oncotic pressure, and urine flow stops. This is the first step in Starling's justification for the glomerulus as a simple filter.

His second step is to reason thus: since urine is normally more concentrated than serum, concentrating processes (he had no idea of their nature) must occur in the tubules as the glomerular filtrate passes along them. He

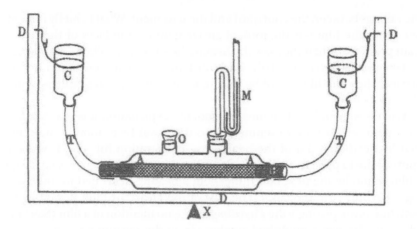

Figure 7-2. Starling's osmometer. The inner compartment (CT) contained a protein-free filtrate of plasma; the outer compartment contained plasma. The two compartments were separated by a semi-permeable membrane. The box (D) was half-filled with water and gently rocked about X (see text). (*Physiological Society; Blackwell Publishing, with permission*)

knew that diuresis (an increase in urine flow) could be produced by injecting solutions of glucose into the circulation. So he injected varying amounts of glucose solution into anesthetized dogs, produced a diuresis of dilute urine, and showed that the greater the dose of glucose injected, the closer does the concentration of urine approach that of serum. Starling reasoned that the glucose produces an increase in the volume of blood ("hydraemic plethora"), which gives rise to an increase in flow through the walls of the glomerulus or capillaries, and an increase in flow through the tubules. He reasoned that this increase doesn't give the absorbing mechanisms time to operate, so the final urine is diluted. All this fits perfectly with the glomerulus as a physical filter, and subsequent research has shown it to be so; it is the complex nature of the processes in the tubules that became the focus of renal physiologists.

One end—the eastern end—of Starling's Institute contained two or three rooms that were the first department of pharmacology in Britain. The department was funded by the American Carnegie Foundation (what would UCL have done without transatlantic generosity?) and the first professor was Arthur Cushny (1866–1926), who started research in his tiny department in 1905.

Cushny, almost an exact contemporary of Starling's, was a Scotsman, and had held the first chairs of pharmacology at Michigan and Johns Hopkins Universities. His relevance here is that he was an authority on the kidney, and Starling, sometime about the beginning of the war, asked Cushny to write a monograph for a Physiological Society series that he was editing. *The*

Secretion of Urine was published in 1917, and became a classic. In it, Cushny reviewed the kidney literature, and the book's preface is actually a letter to Starling: "To Colonel Starling, MD, FRS, etc, British Expeditionary Force. My Dear Starling . . . no other organ has suffered so much from poor work as the kidney, and in no other region of the subject does so much base coin pass as legal tender . . . a view is presented which embraces some of the features of each of its precursors. I have therefore called it 'The Modern View' . . ." (Cushny, 1917).

"The Modern View" is unambiguous about glomerular function, for Cushny is a Ludwigian, supporting his argument with Starling's 1899 paper on the glomerulus. But tubular function is a real problem for Cushny. He knew, for example, that in order to explain the excretion of urea by filtration alone, a large volume of glomerular filtrate must be formed per minute, and most of the water and salts would be reabsorbed as the fluid passes along the tubules. Moreover, he worked out that some substances (glucose for example) only appeared in urine when a certain plasma level was exceeded. Such substances he called "threshold bodies," and he was quite right. But there were other bodies that seemed to have a higher concentration in urine than plasma (sulphate ions and urea for example). At any suggestion that the kidney might actually secrete molecules into the urine, Cushny's credulity fled. He regarded such phenomena as "vitalistic," smacking of Heidenhain. Because Heidenhain had proposed glomerular filtration to be a secretory process— and Cushny disposed of this with Starling's evidence—all renal secretion was anathema to Cushny. But Heidenhain had done an experiment (it had been as long ago as 1874) showing that when the dye indigo carmine was injected into the blood stream, it could be found staining the distal part of the tubules, but not the glomerulus. Cushny tries very hard to demolish Heidenhain's experiments, but doesn't convince us. Any sort of tubular secretion was to have no part in Cushny's "Modern Theory." In spite of its scholarly bigotry, Cushny's book had an important influence on renal physiology and medicine.

Starling, on the other hand, seemed to have little difficulty in accepting tubular secretion (he had worked with Heidenhain, and possibly had less trouble accepting the indigo carmine experiment). But he had other conceptual problems with the kidney. There was no method at this time of measuring the rate of flow of fluid across the glomeruli—the glomerular filtration rate (GFR). Starling estimated the GFR—on the basis of the amount of urea appearing in the urine—as 30 liters/day. Since urinary output was about 2 liters, this meant that 28 liters of fluid must be absorbed by the tubules. This struck Starling as ridiculous, a clumsy and inefficient way of arriving at urinary output ". . . and as we have seen, the occurrence of actual secretion of urea by the cells of the tubules takes away the necessity of assuming any such wasteful proceeding. It is probable that the actual volume of glomerular filtrate in the 24 hours may not exceed to any large extent the actual amount of urine formed by the kidney in this time" (Starling, 1899).

Physiologists look for simplicity and neatness in the body's activities. Here was a situation which was so lacking in either that Starling had to construct his own version of the truth. He only felt comfortable proposing the filtered volume to be the same as the urinary volume. In fact, the truth is even more remote from common sense than Starling believed. For we know that no less than 180 liters of fluid pass across the glomeruli per day, with about 178 liters reabsorbed in the tubules. More than 99% of the water, sodium, and chloride ions that pass across the glomeruli is re-absorbed. Apart from the intrinsic unlikeliness of the arrangement, it represents an extravagant expenditure of energy by the kidney. It is as though, in order to throw away kitchen waste, everything in the kitchen (furniture included) has to be carried to the waste bin, and everything—except the actual waste—carried back into the kitchen.

The American physiologist Homer Smith subsequently proposed an evolutionary explanation for why the kidney should be organized in this strange way (Smith, 1959). Life began in a brackish watery environment and Smith proposed that the inhabitants' body fluids reflected this; they had no reason for being parsimonious with water, sodium, or chloride. They merely had to have some simple conduit to get rid of body fluids containing waste products (such as urea). But the next evolutionary step involved animals moving to fresh water: suddenly there was a need to conserve sodium and chloride (for the body fluids needed them for osmotic stability); the nephron developed absorptive pumps to prevent the loss of what were now precious materials. Finally, the evolutionary step came when some animals moved from fresh water to land. Now the precious materials were sodium, chloride, and water; the tubule's parsimony extended to all three. Smith's evolutionary explanation may or not be true, but it is the best we have.

The Heart—Lung—Kidney Preparation: Basil Verney

Starling felt that there were two possible ways of discovering more about how the kidney worked. One way was to make changes to the blood passing to the kidney in the whole animal, and to look for changes in the urine. (This was clearly the only practical approach in man.) The second way was to remove the organ, perfuse it with blood, accurately controlling the variables involved in the perfusion and examine the urine. Starling's natural bias was toward the second way, because he had, in the heart–lung preparation, a method whereby he could physiologically perfuse the kidney with the animal's own blood.

He had actually proposed using the heart–lung preparation in this way before the war. In 1914 he suggested it to Bainbridge and Evans, working in his institute (it was about a year before they became gas officers under Starling at Millbank). They described the working of the heart–lung–kidney preparation (Bainbridge and Evans, 1914); it involved the use of

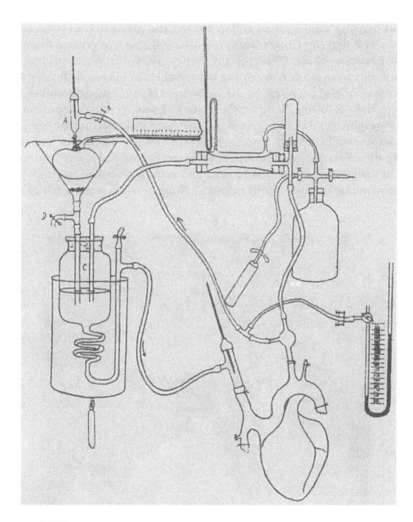

Figure 7-3. The heart–lung–kidney preparation, a diagram that represents half a laboratory's worth of equipment. A heart–lung preparation (bottom right) provided pulsatile, oxygenated blood perfusing the kidney of a second anesthetized animal (top left). Urine secretion was measured along the graduated horizontal tube. Blood pressure was maintained with the Starling resistor (top right). (*Physiological Society; Blackwell Publishing Ltd*)

two anesthetized dogs and a great deal of equipment. Bainbridge and Evans made this complicated experiment work on three occasions; but the kidneys produced virtually no urine, even when perfused at 100 mm Hg. This was disappointing, as the oxygen consumption of the kidneys suggested that they were working normally.

After the war, in 1921, Starling decided that he was going to give the heart–lung–kidney preparation another try. With Basil Verney (a Beit fellow, and one of Starling's most gifted collaborators) the preparation worked. The two showed that the concentration of urea in urine was greater than that of the perfusing blood (Starling and Verney, 1924–25), which proved that the tubules were actively secreting urea. Sulphate ions and the dye phenol red were also secreted by the perfused kidney. The addition of cyanide to the perfusing blood enabled Starling and Verney to confirm the energy-requiring tubular functions that had been proposed at that time. The tubules normally actively absorbed chloride (whose excretion increased under the effect of cyanide); they actively secreted urea (whose concentration in urine decreased with cyanide) and actively absorbed water (so that urinary volume increased with cyanide.) None of this was particularly new;

Figure 7-4. Basil Verney (1894–1967). (*Wolfgang Sievers, Melbourne, with permission*)

it was confirming older work. But there was one way in which Starling and Verney's research really broke new ground. Even in the absence of cyanide, their perfused kidneys secreted urine that was significantly more dilute than the plasma of the perfusing blood. (Normal urine is more concentrated than the plasma supplying the kidney.) They first of all showed that neither dilution of perfusing blood or the addition of sodium chloride to the perfusing blood made any significant differences to the low excretion of sodium and chloride seen in the perfused kidney.

"We suggest therefore, that some substance or substances with an action similar to that of pituitrin are usually present in the intact mammal and serve as the means whereby the kidney is controlled in its important function of regulating the output of water and of chloride." They tested this conclusion by adding a pituitary extract to the perfusing blood, and reduced the water and increased chloride excretion in the urine.

This observation was the spring-board for the rest of Verney's research career. (He subsequently became Professor of Pharmacology at UCL in 1926 and then at Cambridge in 1946.) His next experiment was to show, by Starling-type perfusion experiments, that the factor responsible for urinary concentration originated from the head; then he narrowed the site of origin down to the posterior pituitary gland in the brain; then showed that antidiuretic hormone (ADH, or vasopressin) was released from there. The release was controlled by osmoreceptors (Verney's word) in the supra-optic nuclei of the hypothalamus. An increase in osmotic strength of the blood reaching this part of the brain stimulated the release of the antidiuretic hormone. Nervous influences inhibited the hormone's release. These few sentences represent about twenty-five years of Verney's research life. Verney wrote about his years of research at UCL (1921–1925) "under the direction of Professor Starling, and I cannot exaggerate the greatness of the debt I owe him for the instruction, stimulus, and encouragement he gave me . . . there was intense intellectual activity in the department and the best of good fellowship between those who were working there, and this happy atmosphere was undoubtedly derived from Starling's objectivity, immense vitality, biological insight, and experimental skill" (Daly and Pickford, 1968).

Verney moved away from Starling's use of anesthetized dogs, for anesthesia has significant effects on blood flow to the kidney. Verney's outstanding surgical techniques were learned from Starling, and (using anesthesia) he implanted catheters into his dogs. The actual experiments were then performed with the dogs fully conscious, and because of Verney's care for them, resting quietly on the bench top. On one occasion, when he was giving a demonstration to the Physiological Society, a member wrote: "A dog, cannulated in several places, was lying quietly on her side when a well-known physiologist came in and took a look around, then said: 'Verney, what anaesthetic are you using?' Verney, without replying, leant forward and murmured "Linda" (or whatever the dog's name was) she promptly raised her head, took one contemptuous look at the seeker for information, sighed, and laid her head down again. Verney continued with his demonstration without comment" (Daly and Pickford, 1968).

Starling and Verney's experiments were important, but at that time were not the cutting edge of renal physiology, for this had moved to America. Two developments in particular had taken the physiological momentum across the Atlantic. One was the brilliant use of micropipettes, by A. N. Richards (e.g., Wearn and Richards, 1924) to sample glomerular fluids in the frog's kidney (the glomeruli are close to the surface of the frog's kidney, so are easier to identify); they also perfused the lumen of single tubules, which ultimately led to confirming the various tubular processes that had previously only been inferred. Second was the introduction of the concept of "clearance," which revolutionized thought on the kidney. This was thanks to a distinguished physiological chemist, Donald Van Slyke: ". . . while seeking a solution of how to dispense with mathematics for the benefit of the medical profession, it occurred to him that all that the equation for high urine flow said was that in effect some constant volume of blood was being cleared of urea in each minute . . . in my opinion this work has been more useful to renal physiology than all the equations ever written" (Smith, 1959). Starling witnessed the first micropipette results from America, but he barely lived to see the first measurements of clearance; he would have had to seriously revise his views on glomerular filtration in man when clearance measurements showed it to be 180 liters a day!

The Foulerton Professorship and A. V. Hill

Life was humming for Starling in 1921. Florence notes that if he was not doing an experiment he had a stream of visitors, who would often exhaust him. She persuaded him to walk home to Taviton Street to have lunch in peace, where he had crackers and cheese (his favorite at the time being Roquefort) and a bottle of stout. Experiments—usually heart–lung preparations—were with Verney, Anrep, or Ivan de Burgh Daly, who had joined the department in 1919. Writing to Muriel, he said of Daly "You don't know him—a great enthusiast who wants to hit the stars—I like that—but not very strong in the bow-arm—but, after all, enthusiasm is more important than intellect." "Not very strong in the bow-arm" is hardly flattering, but it didn't prevent Daly from becoming the Professor of Physiology at Birmingham in 1927.

Daly wrote about his experiences in the laboratory: "The atmosphere of the department was light-hearted. There was a feeling that a new era of intellectual freedom had arrived following 'the war which was to end all wars.' We arrived at 9 am, lunched in the old refectory, had tea in the room on the mezzanine floor around Sharpey's table as in the prewar days, and left between 5 and 6 pm. Some of us stayed to work during the night, especially those using galvanometers, because the electrical installations of the department left much to be desired" (Daly, 1967).

Early in 1922, the Royal Society decided to set up one or two research Professorships, for in 1919 they had been left about £70,000 by Miss L. A. Foulerton (the executrix and sole legatee of Dr. John Foulerton) in order to support medical research. The Royal Society set up a committee of six (chaired by the ubiquitous W. B. Hardy) who advertised for applications for research Professorships (£1,400 per year for 5 years in the first instance) in June. The post was advertised worldwide, and there were 30 applicants, five of whom did not even provide the title of their proposed research. Others were simply eccentric: W. W. Strong, of Mechanicsburg, Pennsylvania, wanted to research on "Effects of a magnetic field on optimism and gloom." There were four serious applicants: Joseph Barcroft, John Ledingham, Edward Mellanby, and Starling. Starling saw his Foulerton application as "interfering with a delightful Cambridge scheme! They want A. V. Hill to succeed Langley when he goes [Langley was 70 and died at 73 in 1925, still professor at Cambridge]. But A.V. won't go in against Barcroft, who is waiting for the job—so the Foulerton seems to come as a gift from Heaven. Appoint Barcroft [to the Foulerton] then he is out of the way and they can make A.V.H. succeed Langley. Well you see I am rather formidable competition, so it has thrown confusion into the ranks" (Letter to Muriel Patterson, July 16, 1922).

Starling was appointed. The Royal Society minutes show an odd aspect of the event—on that particular day two extra committee members were present. They were Charles Martin and Grafton Elliot-Smith (the Professor of Anatomy at UCL); both would have been strong supporters of Starling. No votes are recorded, and it does look as though the committee had been suitably adjusted to help Starling. Perhaps it was just a coincidence.

Life as Foulerton Professor would save him from a lot of teaching and administration, and he moved into a new suite of rooms on the third floor of the Institute (one floor higher than his old lab). The money for his new rooms was provided by the Rockefeller foundation.

Archibald Vivian Hill had been Professor of Physiology at Manchester since 1920. (He disliked both his first names, and only ever used his initials.) Hill was neither medically nor physiologically qualified, but his mathematical and physical background made him a new kind of physiologist—he would subsequently be called a biophysicist. Hill had originally asked Starling for advice on "whether he should go [to Manchester], or stay in the pleasant seclusion of Cambridge." Starling's instant advice had been "My dear Hill, you don't know a word of physiology, but I think you ought to go there" (Hill, 1962). Hill had made a success of Manchester, and now, in 1923, Starling was plotting that Hill should succeed him as Jodrell Professor. "A full vigorous personage—35 years old—who will be the most important man in the physiological world, and likes to be in the scrum. A good fighter! Foster [Gregory Foster, Provost of UCL] had heard about him and asked me. I said yes, I think he's the right man. But of course, he's not a medical man, he is not even a physiologist. Poor old Foster—it was as if the bottom had dropped out of his universe. He gasped 'But he is a professor of physiology.' I said 'oh yes, and he will do very well as head of this institute'" (Starling, 1922).

In this discursive letter, Starling moves to a related subject: "The mistake they make at Cambridge is not the omission of a medical training—but letting a man be a physiologist only—a man should always approach our goddess bearing gifts—it may be medicine—with its many-sided training in science and mankind—or it may be physics, or zoology—or chemistry—I must regret that Leonard Bayliss [Starling's nephew] did not take his tripos [three-part degree] in something else before turning to physiology." Perhaps this was because his father William had done it that way.

> Next Friday A. V. Hill (and possibly his missus) come and stay the
> whole week with us. I have asked him to dine at the Athanaeum
> to meet Bragg, Donnan, Elliot, and Boycott [many of the scientific elite of UCL] . . . Bayliss will dine him and all my staff at St
> Cuthberts. We shall be a strong team when he comes . . . So, as I
> told Foster, 'the best thing I have done for the college will have
> been to resign . . .' (Starling, 1922)

His resignation from the Jodrell professorship meant that he could give all his time to research. One of the first things he did was to chair Hill's inaugural lecture. In the lecture, Hill recalled how Starling told him he knew no physiology, but that he should certainly take the job at Manchester. Now Starling had persuaded him (still in his uneducated state), to come to UCL (Hill, 1923).

A few months later, Hill was awarded a Nobel prize for his work on muscle. The students carried him shoulder-high round the college, and took him to Starling's laboratory in the top of the Institute. They loudly demanded: "Who says he doesn't know a word of physiology?" Starling replied "I did— he doesn't know a damned word." And the students took Hill away (Hill, 1923). The story has another side to it, for Starling had been the subject of serious discussion by the Nobel Committee—which he must have been aware of—and his reply could have been seen as a hint of sour grapes. (We will return to Starling and the Nobel committee in Chapter 8.) But bitterness was not part of his nature, and the relationship between the two men seems to have been one of mutual admiration and affection.

Florence's Leg

In the early nineteen-twenties, horse buses operated up and down Gower Street and Tottenham Court Road. Toward the end of June 1921, a few months after moving to Taviton Street, Florence was hit by a horse bus while riding her bicycle (Henderson, 2000). She broke no bones, but had a large area of skin avulsed from her right thigh. She was taken to UCH, where she was seen by Wilfred Trotter, the hospital's senior surgeon. Perhaps because Florence was an important patient, Trotter consulted Sir Arbuthnot Lane at Guy's (who had operated on Ernest's colon the previous year). Plastic surgery hardly existed at this time, and Trotter and Lane contemplated amputating Florence's leg, for there was no way that new skin could grow from the edges of the wound.

Florence was discharged home, and Ernest became her nurse. He wrote to Muriel:

> At present, she eats and sleeps well, has good colour and is
> cheerful, in spite of having 60 square inches of granulating skin. I
> go up every morning and dress it, before going to the lab. When
> term starts I shall have to do it in the afternoon . . . On Arbuthnot
> Lane's advice. I asked Gillies to see her [Gillies was the rising star
> of plastic surgery in England at that time] He thinks he can close—
> or nearly close—the wound with skin (not grafts, but whole skin)
> from the abdomen and the other thigh.

Gillies began a series of five operations. The first was to construct a 12 inch by 3 inch skin sausage (a pedicle) from Florence's umbilicus to the top of her thigh, with the pedicle receiving a blood supply for both ends. After 2 or 3 weeks, the lower end of the pedicle was partly detached and the bottom 3–4 inches opened out and sutured onto the raw area. Since the pedicle was still attached at her umbilicus, Florence had to keep her hip and knee bent all the time (illustrated by Ernest in letters to Muriel). She was very stoical, and was helped by a stream of visitors. Trotter himself came every day and dressed her wound. A night nurse (Nurse Mcintyre) was provided to stop Florence from straightening her leg in her sleep. Christmas 1921 saw a family party in her room in UCH.

Gillies lengthened the pedicle by 3 inches at her umbilicus, making it 15 inches, believed to be a world record for the time. This left her more comfortable, but the surgical wound around her umbilicus did not heal satisfactorily. She was moved to a private nursing home, and wrote, with her usual eye for detail:

> As the stomach wound has not healed by first intention (Gillies
> owns he made a mistake to put collodion on in the operating
> theatre so that any little accumulation had to force its way
> through) it is now beginning to granulate, and that of course is a
> longer process. Kilner (Gillies' assistant) took out the stitches
> today (8 days after the op) an extra long time, owing to the above
> cause. He is evidently musical, as he asked to put on a record
> when he came in, and another after he had finished (did I

mention that Ernest had brought round his 37/6d German
gramophone the other day?). It was the first and last movements
of the Kreutzer Sonata, violin part by Marjorie Heywood, and
Sister stood and listened too—the whirr is practically nothing . . .
(Letter to Muriel Patterson, January 4, 1922)

It paints a memorable picture of life in a private nursing home in the
nineteen-twenties; no room was complete without a 37/6d German gramophone, with Beethoven to accompany a little minor surgery.

The pedicle was rotated and resulted in Florence's umbilicus being sited
on her right knee. The area was totally covered by graft. Subsequently, whenever she was sitting down, Florence had to put her leg up on a chair. (Perhaps this large area of transplanted skin was inadequately provided with new
lymphatic vessels, which would have caused it to swell.) Gillies was so proud
of his achievement that when asked by the journal *The Practitioner*, in 1956,
for an account of his most memorable case, he chose Florence and her leg,
34 years before (Gillies, 1956). Although ultimately her leg was saved, Florence probably never again danced with her husband.

Pavlov, Starling, and Dialectical Materialism

The Bolshevik revolution of 1917 had thrown Soviet science into turmoil.
I. P. Pavlov, as a grand old man of Russian science—he was 70 in 1919, but
still active in research—seemed particularly vulnerable. In 1921, Abraham
Flexner, working for the Rockefeller Foundation, and concerned about
Pavlov's welfare, wrote to Starling (Flexner, 1921). The Foundation wanted
to support Pavlov and his family in America but Flexner was sure that they
would not like to live there. Could Starling find support for Pavlov in England? (Flexner actually suggested that Pavlov should go to Dale's MRC in
Hampstead.) The Rockefeller Foundation was prepared to fund Pavlov in
England, but only for a second year; the first year would have to come from
an English source.

Starling first approached Hardy, the secretary of the Royal Society, proposing £600 as a year's support for the Pavlov family. ". . . If this sum could
be assumed for the year from the time that Pavlov lands in England, we could
put his friends onto the job of smuggling him out of the country. He is not
an easy man to smuggle . . . his daughter would probably be able to get some
work in the course of the next two years and could add to the family income."
But Starling's quixotic plan didn't appeal to the Council of the Royal Society. He then wrote to the secretary of the MRC, Walter Fletcher, who was
just as unhelpful: "We must clearly distinguish between philanthropy on one
hand and financing research on another. The business of the Russian men
is a deplorable tragedy; we have three other applications from Russia before

us now, all grave cases, but offering technical difficulties—believe me—yours sincerely, W. M. Fletcher."

Bayliss had his own approach to Pavlov's position—it was rather less ambitious than Starling's. Bayliss wrote to Hardy, suggesting that Pavlov should visit England and give a series of lectures on conditioned reflexes. "I take it some funds would be available for his journey. I will ask Arrhenius [the Danish physical chemist] to help." Bayliss's plan was as impractical as Starling's, for Pavlov spoke no English, and lectured "in a German of his own invention" (Evans, 1964).

In fact, Pavlov never came to England after this time, for in Russia his luck was changing. The government acknowledged that science could be a power for good, providing that scientists reasoned in accordance with dialectical materialism. Conditioned reflexes seemed to fit this bill, and suddenly Pavlov could do no wrong. In 1921 three commissars (Lenin, Gorbumov, and Gliasser) issued a statement that Pavlov should be given exceptional treatment: a large new laboratory was to be built, his family provided with a new luxurious flat and their food rations doubled (Babkin, 1949). So the concerns felt by the west for this Grand Old Man vanished overnight, and Pavlov lived comfortably in Russia until his death in 1936.

Starling and Bayliss felt that the scientific world had been deprived of Pavlov's findings since 1904, when his research had switched so suddenly from gastroenterology to conditioned reflexes. (It has been argued that the switch was a direct consequence of the discovery of secretin by the London pair: Todes, 2002.) Pavlov would have had to repeat many of his gastroenterological experiments, taking account of hormones. Virtually none of the conditioned reflex work had been published in English. Working in Starling's laboratory was the one person who could translate Pavlov's work perfectly—Gleb Anrep, who was not only a pupil of Pavlov, but also bilingual in Russian and English. Starling persuaded the Royal Society to pay Anrep £100 for translating Pavlov's twenty-five lectures on conditioned reflexes. Anrep did so, and in 1927 they were published by Oxford University Press (Pavlov, 1927); it was an exceptional feat of translation, for many of the Russian words were actually new to science. Thus we owe the use of the word "conditioned" to Anrep.

Starling believed at this time that conditioned reflexes provided the way forward in studying the central nervous system. He wrote an adulatory biography of Pavlov in *Nature* in 1925 (it was one of a series unpromisingly titled "Scientific Worthies"). In his generous and sporting way, he makes no mention of the fact that Pavlov failed to acknowledge secretin or hormones in his Nobel Prize acceptance speech. Nor does Starling even hint at the reasons for the turnaround that occurred in Pavlov's research in 1904. Starling's admiration for conditioned reflexes as a tool for studying the cerebral cortex rested on the fact that the techniques involved trained, unanesthetized animals. Conventional neurophysiological techniques, Starling argued, suffered seriously from the necessity for anesthesia and the surgery needed to expose the brain (Starling, 1925). But he was wrong, for after Pavlov's death it became more and more apparent that conditioned reflexes taught us little about the real workings of the brain. Study of them lasted longer in Russia than anywhere else, because belief in their usefulness was still in accord with dialectical materialism, while at the same time Russian science was somehow paying homage to its greatest physiologist.

The Harveian Oration: Hormones Revisited

The Royal College of Physicians invited Starling to give the 1923 Harveian Oration—a notable honor. He had been elected to fellowship of the college, but, unlike other Harveian orators, was not a practicing doctor. Even so he was an apt choice, for many felt that he had achieved as much in his research on the heart and circulation as anyone since Harvey. Starling used the occasion to spread himself, for the printed lecture is about 8000 words long and must have severely taxed the concentration of the fellows. It was actually published in both national medical journals (the *Lancet* and the *British Medical Journal*) in full, on the same day. (Did their editors not speak to each other?)

The oration's title, "The Wisdom of the Body" is biblical. (Who hath put wisdom in the inward parts? Or who hath given understanding to the heart?— Job 18:24.) It was a title later borrowed by Walter Cannon for his well-known book on homeostasis (Cannon, 1932). In his preface, Cannon observes, rather intriguingly: "The late Professor Starling, of University College London, gave the Harveian Oration before the Royal College . . ."

Starling's lecture begins with a tribute to Harvey. "His treatise on the motion of the heart is throughout so modern in spirit, so akin in conception and treatment to records of research in the present day, that we may easily fail to appreciate the stupendous advance in human physiology that it embodies, and may wonder that we have had to wait so long for the full fruition of Harvey's discovery." (Is Starling implying that *his* work is the full fruition of Harvey's discovery?) Harvey had observed that the output of the heart varied under different circumstances, and Starling uses this to develop his own theme, the adaptive power of the organ, that is, the Law of the Heart. He reviews his own findings, quoting Blix and A. V. Hill (but not, interestingly, Frank); he was "seeking a complete description of the acts of excitation and contraction as molecular events occurring at surfaces."

Starling uses the heart as a stepping stone to hormones, his main theme. Harvey referred to the "heart as the sovereign king of the circulation, throughout he continually recurs [sic] to what we would now describe as 'the integrative action of this organ.' In virtue of the circulation which it maintains, all parts of the body are bathed in a common medium from which each cell can pick up whatever it requires for its needs . . . changes in any one organ may therefore affect the nutrition and function of all other organs, which are thus all members one of another." In this elegant way, Starling moves to his main theme.

"It is now 18 years since I first drew the attention of this college to the chemical messengers or hormones, employed by the body for this purpose" (the Croonian Lectures of 1905, when he introduced the concept of "hormone"). He contemplates the knowledge of the substances that has accumulated since then: insulin, sex hormones, and thyroid hormones. (Thyroid hormone had recently been shown to be an iodine derivative of tryptophane.)

"It seems almost a fairy tale that such widespread results, affecting every aspect of a man's life, should be conditioned by the presence or absence in the body of infinitesimal quantities of a substance which by its formula does not seem to stand out from the thousands of other substances with which organic chemistry has made us familiar." Such thoughts seem ingenuous to us now, for they are taken for granted by receptor theory. He also reviewed the early work of Schäfer on the secretions of the adrenal medulla, and provides a paragraph on the Austrian, Paul Kammerer. Kammerer claimed that the pituitary melanocyte-stimulating hormone could affect germ cells. In this way salamanders brought up against dark backgrounds tended to produce offspring with dark skins, and Starling was naturally wary of such Lamarckian heresy. Later, when it seemed that his findings had been fabricated, Kammerer shot himself (Koestler, 1971).

The recent isolation of insulin by Banting and Best in 1922 confirmed Starling in his belief that hormones were going to be of the greatest clinical importance, and that we were on the threshold of "an hormonic revolution." This inspired him to construct a framework for the future clinical use of hormones. "There seem to be three possible methods by which we medical men can interpose our art in the hormonic workings of the body [even Starling had his pompous moments!]:

1. Discover the effective stimulus for the secretion of the hormone, and so exert control of its secretion.
2. Where a disorder is known to be the result of diminished production of a hormone, we may extract the hormone from the corresponding gland in animals . . . we may look forward to the day when the chemical constitution of all these hormones will be known, when it will be possible to synthesize them in any desired quantity.
3. The ideal, but not, I venture to assert, the unattainable, method will be to control, by promotion or suppression, the growth of the cells whose function it is to form these specific hormones."

In these few sentences he anticipated the next 80 years of endocrinology. He finished in his best aphoristic mode, on the beauty of science: "The strife of tongues is impossible, for science throughout the world has but one language, that of quantity, and but one argument, that of experiment."

The Death of William Bayliss

In 1923 the Institute of Physiology was very full of researchers. Not only had Hill bought a new collection of workers with him (notably C. N. H. Long and H. Lupton) from Manchester, but biochemical research was flourishing. The department changed its name to "Physiology and Biochemistry." Most of the biochemistry was concerned with vitamins, under the leadership of Jack Drummond, who was the recently appointed Professor of Biochemistry. Today, Drummond's biochemistry would be called "nutrition," and he was to write *The Englishman's Food*, a classic of that subject (Drummond and Wilbraham, 1939). But nutrition as a formal academic discipline was not yet born, being still part of biochemistry. University College was actually very advanced in facilitating the splitting away of chemical subjects from physiology, and William Bayliss had been there at the start, with his renowned *Principles of Physiology*, first published in 1915.

Bayliss showed signs of illness in 1923. We know nothing of the nature of his complaints, but they must have been serious, for a new edition of his *Principles* was due and it was clear that he was not going to be able to finish it. Starling and Hill divided the chapters up among twenty-three colleagues, and, as a gesture of affection for their friend, each author had the task of updating a chapter.

Sadly, Bayliss died before the labor of love was completed. He died in August 1924, at age 64, at St. Cuthberts, where so many of the people in this book spent happy hours. His death certificate gives the cause of death as "lymphadenoma"—an old-fashioned word for lymphoma. (Since he had been ill for less than two years, it would seem likely that he died of Hodgkin's disease, or some related malignancy.) Of the obituaries written about Bayliss, several were by Starling, who wrote two of them for German journals. The most heartfelt was one that he wrote for the magazine of University College, where Bayliss had spent his whole professional life. When the edition of the *Principles* appeared, the authors all signed a single copy, inscribed it, and presented it to Gertrude, who was very moved. Leonard Bayliss, William and Gertrude's son, was working with Starling at the time of William's death, and went on to produce, in 1959, a revised edition of his father's magnum opus. (Was there ever a more family-oriented laboratory than Starling's Institute?)

Joseph Barcroft (1924) wrote: "Bayliss delighted in the society of other scientific men: not least that of young physiologists, many of whom gravitated to the Institute at University College in order to work with Starling and himself. Bayliss's honesty and generosity of outlook, the simplicity and nobility of his mind, his faculty of getting to the bottom of problems, coupled with his great erudition, made conversation with him a pleasure and an education."

8

The End of the Trail

The Pattersons Return

Muriel and Sydney Patterson had gone to Australia in 1919, and their son Tom was born there in 1920. But they were not happy—partly because Sydney was not a natural administrator, and partly because Muriel, being literary and musical, found the place lacking in culture. (Sydney was known as "Pat" in the family correspondence, thereby avoiding confusion with any Australian cities.) Back in England, Ernest was constantly trying to lure them home, and his letters are full of madcap schemes—joint holidays halfway between England and Australia, or chairs of medicine or pathology for Pat at miscellaneous British institutions.

Ernest distributed Pat's curriculum vitae enthusiastically, and managed to get him on a short list—of two—for the chair of medicine at Cardiff. Florence, writing from her hospital bed in 1921 (after her leg accident), was chatty to Muriel about the Cardiff post. Florence had a knack of providing information that had clearly originated from Ernest, but had somehow been taken over by her. Thus, in discussing the post at Cardiff: "only two names had been considered—Pat's and Wiltshire (not a local man). Latter is, I believe, a KCH physician, with no research to his name. I gather Wiltshire has a pull, as he is 'interviewable.'" (In sharp contrast to Pat, who was far

Figure 8-1. Sydney Patterson (1891–1960). A Starling collaborator, and married to Starling's daughter Muriel. (*Family collection*)

from interviewable.) Enquiring how things were going, Starling wrote to the Principal of Cardiff, who responded by postponing the whole thing for a year. Perhaps he did not like candidates for chairs with pushy fathers-in-law.

There were unsuccessful short lists for chairs at Guys, Manchester, and Cambridge. Starling must have been very persuasive, for what selection committee would consider appointing a professor who couldn't be interviewed? Finally, Ernest struck gold, though not with a university post. In February 1923, he was corresponding with Edmund Spriggs, who had been a physiologist and physician at Guy's Hospital. Spriggs was running a research-oriented private hospital in a picturesque castle at Ruthin (pronounced "Rithin"), in North Wales. (Nowadays the words "research" and "private" would be mutually exclusive.) Starling sent Pat's C.V. and a photograph to Spriggs. Here is Spriggs's remarkable reply:

Dear Starling

I return the photograph with thanks. I don't suppose my fellow directors would agree to take a man they hadn't, or I hadn't, seen, but supposing for a moment they would, am I right in thinking that Patterson could not be got (if he were willing) immediately, ie April at latest? I see he has done three full years at Melbourne.

In any case may I congratulate you on your son-in-law? There is a ring of enthusiasm in all these testimonials. I should think he would make a most charming and exceptional colleague

Yours sincerely

Edmund Spriggs

The Pattersons needed no more encouragement. Pat gave in his notice at the Walter and Eliza Hall Institute; their Melbourne house was auctioned on May 1st and they were in England by June. They moved into Ruthin Castle in July. Ernest and Florence were naturally delighted, and they had their first sight of grandson Tom.

Ruthin was a far cry from the university chairs that Ernest had been planning, but with over fifty private beds (and some research) it suited Pat well enough. He spent the rest of his career there, doing gastroenterological research while enjoying the life of a country gentleman. He succeeded Spriggs as Superintendent in 1944, and became President of the British Gastro-enterological Society in 1954 (O'Connor, 1991). Four more Starling grandchildren (Jocelyn, Ann, Daphne, and Rowena) were born in Ruthin Castle, and Pat died there in 1960, aged 78.

The Pancreas and Diabetes Mellitus: A Rare Failure

In 1912 and 1913, Franklin P. Knowlton, from Syracuse, New York, came to work in England. In his first year he experimented on the kidney in the department of physiology at Cambridge, and in the second year came to work with Starling at the new Institute (as we saw in Chapter 4). During this year the two men worked on the heart–lung preparation, publishing the first important paper on the law of the heart (Knowlton and Starling, 1912a). Starling's active mind was still playing about with the idea of hormones, and more specifically the relationship between the pancreas and diabetes mellitus. Following the demonstration by Minkowski in Germany (in 1889) that removal of the pancreas produced diabetes mellitus, a flood of research followed: an estimate was made in 1910 that up to that time, 400 publications on the pancreas and diabetes had been published. The problem was to make an extract of the pancreas which would reverse the changes of diabetes. This was usually assessed by the reduction of sugar in the urine. A more sophisticated approach was to measure the concentration of sugar (glucose) in blood—an effective pancreatic extract producing a measurable fall in blood sugar level. At this time, the measurement of sugar in blood needed about 20 mL of blood, which was a serious handicap; by the early 1920s, when insulin was isolated, techniques had so improved that the sugar concentration could be measured in 0.2 mL of blood.

The word "insulin" was introduced by Starling's predecessor, Edward Schäfer, in 1910. (It is extraordinary how many physiological pies Schäfer had his finger in.) He introduced the name, convinced that the substance *had* to exist. He knew that it was produced by the pancreas, because removal of that organ produced diabetes mellitus, and it was a product of the islets ("insulae"), because diabetic patients showed pathological changes in those particular groups of cells.

Starling realized that the heart–lung preparation was a possible tool for examining the effects of pancreatic extracts on the uptake of sugar by the heart. Did the heart from a diabetic dog take up less sugar than a normal organ? He and Knowlton began by measuring the sugar uptake of normal hearts (Knowlton and Starling, 1912b). They added small amounts of glucose to the blood of the heart–lung preparation, and measured its disappearance over a period of one hour. The uptake of glucose in the heart was about 4 milligrams (mg) of glucose per gram of heart per hour. The next step was to make dogs diabetic by removing their pancreas and, a few days after the operation, measure the uptake of glucose by the heart from the blood of that dog. There was virtually no uptake of glucose by the diabetic heart—in five experiments, this range was 0–1 mg of glucose per gram of heart per hour. The next experiment was to see whether a normal heart perfused with diabetic blood would take up glucose. They only did this experiment once, but the results satisfied them: in the first hour of perfusion the uptake of glucose was 3.5, in the second hour 2.5, and in the third hour 1.7. In Starling's words: "The obvious interpretation of these experiments is that the tissues of the body normally contain some substance, the presence of which is essential for the direct utilization of sugar by the tissues. This substance is gradually used up in the tissues, and therefore has to be continually replaced from the blood if the utilization of sugar is to continue." (Notice that Starling doesn't use the word "insulin," in spite of its introduction two years before.)

Finally he asks whether pancreatic extract is capable of reversing the poor uptake of glucose seen in the diabetic heart. He and Knowlton ground up a pancreas, boiling it in acid Ringer's solution, filtering it and then neutralizing the filtrate with a few drops of sodium carbonate solution. This was added to the blood circulating in a heart-lung preparation. The experiment involved measuring the glucose uptake of a diabetic heart perfused with diabetic blood for one hour, then adding the pancreatic extract and measuring the glucose uptake for the second hour. The experiment was done on four occasions, with the first hour's uptake being 1.5, 0.5, 0.5, and 0.5 mg of glucose per gram of heart; the second hour's uptakes were 4.3, 3.0, 2.8, and 3.6 mg of glucose per gram of heart. (The reader was expected to judge for himself how different these two sets of figures were; statistics had not yet appeared in the presentation of results.)

There is no doubt that these experiments show the effect of a pancreatic extract on the heart of a diabetic animal. For the time (1912), the experiments were outstanding, for they were examining the effect of the putative hormone on the sugar uptake of a single organ. They were not estimating the influence of pancreatic extracts on whole animals, as most experiments did, which often only examined the effects of extracts on the amount of glucose appearing in urine.

Then something curious happened. The obvious next step for Starling to take would be to purify and/or isolate the factor in the pancreatic extracts. But he didn't do this: with Sydney Patterson (in his first year of the Beit Fellowships), an expert heart–lung perfuser, he actually repeated the experiments that he and Knowlton had done (Knowlton had returned to New York by this time). Uncertainty about aspects of the paper with Knowlton is expressed: the accuracy of the blood glucose measurements doubted; significant differences in uptake between diabetic and normal hearts are questioned; no estimates had previously been made of glycogen present in the heart, and it seemed likely that diabetic hearts contained more glycogen than normal hearts. Could this glycogen complicate the findings by releasing glucose during the experiment?

It is an experimental scenario with too many variables, and it is possible that Knowlton and Starling were just lucky to get the results they did. It was beginners' luck in a new field. It is possible to interpret these experiments in a totally different way; here is Jay Tepperman (1988): "Sadly another investigator [Patterson] who was asked by Starling to check on Knowlton's work, had difficulty in confirming it. Starling therefore lost confidence in Knowlton's results and publicly repudiated them. Knowlton, a casualty of his own professor's enormous authority, is not even remembered by some authors as they list investigators who almost discovered insulin." The second paper (with Patterson) found serious flaws in the first paper. It is not a fair description of the findings to say that Starling "publicly repudiated" them. Starling's name was on both papers, so was he repudiating himself? It is reasonable, though, to think of Knowlton as a victim of Starling's "enormous authority."

In the early months of 1923, while negotiating with Spriggs at Ruthin over a possible job for Pat, Starling was also writing to him about the first clinical use of insulin in Britain. (Banting and Best had famously described their isolation of the hormone in 1921, and Banting received the Nobel Prize for the work in 1923.) Here is part of a letter Ernest wrote to Muriel (January 31, 1923)

> ... I am asking Spriggs to give me details as to scientific equipment
> and assistants [at Ruthin Castle]. They are just starting to make
> insulin and then to try it out on their new diabetics [This is
> remarkable, though we have no idea of how successful Spriggs'
> venture was]. We are making insulin at UC—have just got the first
> batch (This is secret—every diabetic in the kingdom is clamoring
> for it, and we can't make enough yet.) It is amazing stuff—Pat and
> I missed it—I suppose through lack of faith—9mg to a normal
> rabbit brought blood sugar down .04 and then to .00 and convul-
> sions. Then 5 grams of glucose in 40 cc water subcutaneously and
> in about 5 minutes sat up and hopped away—perfectly well.

Starling and Henry Dale planned a meeting of physicians and physiologists to review the status of insulin, to be held in June 1923; it was held at
UCL, though no records of the meeting have survived. Starling spoke on
the "Theory of action of Insulin." Dale and J. H. Burn reported on the stimulatory effect of insulin on glucose uptake on the isolated rabbit heart; this was
a simpler version of the experiments that Starling had done with Knowlton
and Patterson years before.

The eleventh International Congress of Physiology was held in Tubingen
in 1923; insulin dominated the program. Starling and Dale both intended
to go, but Dale was prevented at the last minute. Starling later wrote this
laconic comment on the German economy, enclosing a letter to Dale from
a third party (October, 1923).

> Dear Dale,
>   This letter was given me by somebody in Tubingen—I forget
> whom. I don't know why he didn't send it—perhaps he had not
> got the million or so marks requisite for its postage.
>   Yours very truly
>   Ernest H Starling

Although Starling was greatly interested in insulin, it was one of the few
of his research topics to which he failed to make a significant contribution. He later told Lovatt Evans that he believed the Knowlton experiments
(rather than the Patterson ones) to be valid (Evans 1964). But by then it
was all too late, for the world had moved on.

The Control of Blood Pressure: The Last Work with Anrep

The mechanisms controlling arterial blood pressure have long been a source
of curiosity to physiologists. Until about 1920, it was generally believed that
the control was a central process, being organized by nerve cells in the brain.
These cells reacted to changes in the pressure of blood reaching the brain

by sending out messages controlling the diameter of blood vessels and the heart rate. Such responses countered the raised or lowered pressure detected by the brain. In this way blood pressure was kept more or less constant.

But there were experiments in the literature that didn't fit with this "central" explanation. The German physiologist Ludwig had proposed that stimulation of a small segment of the vagus nerve (passing to the brain from the large arteries leaving the heart) gave rise to a fall in blood pressure throughout the animal (Ludwig and Cyon, 1866). This small part of the vagus became known as the "depressor" nerve, because it depressed blood pressure when it was stimulated. So it was possible that there existed another sensor for blood pressure levels; it would be a "peripheral" detection device, present in the walls of either the aorta or the carotid arteries. It would make a circuit: blood vessel → nerve → brain → nerve → blood vessel, and was considered to be a reflex.

In the early 1920s, laboratories were investigating this problem in at least three countries The first was Starling's group in London. Starling and Anrep realized that to detect an effect of a pressure receptor they needed to control the circulation in the head separately from the rest of the body. By using a heart–lung preparation, they provided the head of an anesthetized dog with a blood supply that was independent of the rest of its body. The blood supply to the rest of the body was provided by the dog's own heart. The idea behind the work (which was done in 1924, and published in the *Proceedings of the Royal Society* in 1925) was to prove that both central and reflex mechanisms contributed to the control of blood pressure. In the summary of this paper we find:

> A rise of pressure in the aorta causes a reflex dilatation of the
> vessels of the head and neck, i.e. general vasodilatation. A fall of
> blood pressure in the aorta may cause a constriction. These
> effects are carried out through the vagus and are abolished on
> section of both vagi. (Anrep and Starling, 1925)

This looks like an unambiguous description of a reflex mechanism. Yet Starling seems oddly unimpressed by the finding; for he asks no questions about where the nerve endings might be, or the relative importance of central and reflex mechanisms.

In Europe other groups were moving past the UCL findings. In Germany, E. Koch and H. E. Hering drew attention to a swelling in the carotid artery (the "carotid sinus"); stretching the walls of this sinus by raising the pressure within the artery sent impulses via the depressor nerve to the brain. There, the blood pressure centers were stimulated to produce a fall in blood pressure throughout the body. Hering called this arrangement the carotid sinus reflex ("Die Karotissinusreflexe"); he was doing the research at about the same time as Starling and Anrep. Hering's reflex does not appear in the literature until 1927, with Anrep and Starling's paper appearing in 1925.

Figure 8-2. Gleb Anrep (1891–1955). Pupil of Pavlov and Starling collaborator. (*Family collection*)

But Hering gets the credit, probably because he described the carotid sinus sensor and provided a title. He made the subject into a scientifically satisfying story.

Meanwhile, at Ghent, in Belgium, a remarkable father-and-son team was exploring the same world. J.-F. and Corneille Heymans confirmed Koch and Hering's results (Heymans and Heymans, 1926); then Heymans (Jr.) found that perfusing the carotid sinus with certain solutions caused not only a change in blood pressure, but a change in the animal's breathing. They found that close to the carotid sinus was a little button of tissue ("the carotid body"), which was exquisitely sensitive to chemical changes in blood—especially its oxygen content. This is part of a chemoreceptor system that is analogous to the carotid body reflex, and plays a key role in controlling respiration. It was an important finding, and Corneille Heymans was awarded a Nobel Prize for the discovery. It has been speculated that father and son should have shared the prize, but father died before the award was made, and the prize cannot be awarded posthumously.

In about 1922, Corneille Heymans came to work in Starling's laboratory: we have no idea how long he stayed. In an autobiographical essay, he wrote:

> de Burgh Daly and Anrep were active in his department, and
> nearly every day made a heart–lung preparation with Starling
> . . . I shall never forget Starling's outstanding way of thinking.
> One of the best and most interesting moments of the day was at
> tea time when Starling would develop for the group his opinions
> on the progress of his experiments, and giving many ideas
> concerning problems in the field of cardiovascular physiology.
> Starling was, indeed, a very great man and physiologist. (See
> Schaepdryver, 1973)

There is a peculiar irony about this; for Starling was shortly to miss out on the key observations (especially the carotid sinus reflex) that ultimately led Heymans to a Nobel Prize. By 1925, Starling's health was beginning to deteriorate, and although he was still supported by the Foulerton research money, the number of projects and collaborators was falling year by year. Anrep continued with the control of blood pressure experiments, but without the collaboration of his mentor. According to F. R. Winton (who succeeded Verney as Professor of Pharmacology at UCL) Starling was "too ill" to finish this series of experiments. They made up the last piece of sustained research of his life.

Foulerton (2)

At about this time, the Royal Society's Foulerton Research Committee, finding itself surprisingly prosperous, decided that it could afford another research professor. After advertising, the committee decided on Joseph Barcroft from Cambridge, and in July 1925, offered him the job at £1400 a year (tenable at Cambridge). Barcroft was approached secretly, however, because he turned the Foulerton offer down, in spite of having applied for it. It seems likely that he was promised the Cambridge physiology chair, for a year later he became professor there on the death of Langley.

The Committee, in the gung-ho style of such bodies, decided that they were going to offer the second Foulerton professorship to someone who hadn't actually applied: A. V. Hill. (To him that hath shall be given.) Hill accepted the offer gratefully, for he saw it as a way of avoiding all the fiddly bits of administration of the Jodrell chair. Not quite every bit, for the Foulerton committee specified that Hill remained a member of all the university boards on which he sat, a condition that they had not made for Starling. This could have been a tacit appreciation of Starling's failing health.

We have two clear glimpses of Starling's physical state at this time. The first is from Maurice Visscher, a distinguished American physiologist who later became President of the American Physiological Society. He had worked with Starling in 1925–26, and recalled: "He had told me that . . . he was experiencing a recurrence [of his disease] but he refrained from talking about it, and instead went right ahead with his research and writing programme. He was a man of extremely simple tastes and extremely straight-forward in his contacts with students and associates" (Chapman, 1962).

The second glimpse, rather more startling, comes from Sir Andrew Huxley, who became Jodrell Professor 33 years after Starling's death. Sir Andrew writes: "When I arrived at UCL in 1960, the head technician in the department was Charlie Evans, who had started working there when Star-ling was still head of the department. There was an ante-room on the way from the professor's office to the lecture room and Evans told me that this was built because 'Starling used to fart so loud that it could be heard through-out the lecture room.'"

There is a reasonable explanation for Starling's delicate problem. Nor-mally, the contents of the small intestine pass intermittently into the colon, this being controlled by the ileo-colic valve. If, during the removal of half of his colon in 1920, the ileo-colic valve had been damaged, small intestinal contents would pass continuously into the colon. Such "dumping" would give rise to large amounts of gas. We have no more evidence as to how much this problem interfered with his everyday life, but if it led to the building of a special anteroom, it suggests that it was seriously debilitating. The origi-nal (1909) plans of the laboratory show a small room next to the lecture theatre labeled "dark room"; and such a room is there today.

Starling's department now had two important "extra" professors, and a new Jodrell Professor was needed. It was no great revelation when Lovatt Evans was appointed; he was a Starling protégé through and through, and almost every piece of research he did grew out of work done with Starling. Thus Evans's most significant contribution to physiology was the elucidation of the metabolic fuels used by the heart, and he made use of the heart–lung preparation to achieve this. He might have been daunted having Starling and Hill looking over his shoulder, but the three were old friends. In the Royal Society's Biographical Memoir of Lovatt Evans we read:

> This was the first time that Starling, Hill, and Evans found
> themselves in the same department, but not the first time they
> had an intimate scientific relationship. Just before the First World
> War, Patterson, Piper and Starling obtained evidence that the
> force of muscular contraction was determined by the diastolic
> length of the cardiac fibres—Starlings Law of the Heart. The
> work of Evans on the oxygen usage of the heart had pointed in

the same direction. Starling thought that perhaps this was a property common to all muscle, and sent Evans to Cambridge to study (under Hill) the relation between length of fibre in skeletal muscle and the heart produced on isometric contraction. Their results confirmed his expectation . . . heat production in skeletal muscle, like the force of systole and the oxygen usage of the heart, reached a maximum as stretching progressed, and then declined. Starling believed that this was what happened in heart failure. (Daly, 1970)

The Nobel Prize

Starling was first proposed for a Nobel Prize in Physiology or Medicine in 1913, by Otto Loewi, from Graz, in Austria (who himself was to win a Nobel Prize in 1936) and E. Lahousse, from Ghent in Belgium. The Swedish documents, which have been since released by the Nobel Committee, are headed: "*1913: secret document ('Secret Handling'). Investigations involving Schäfer, Bayliss, and Starling*" (Nobel committee, 1914).

The assessor for the proposals was Professor J. E. Johansson, a Swedish physiologist who exerted a powerful and eccentric influence over these prizes for more than twenty years. He began his report by removing Bayliss's name from the candidates. He justified this by pointing out that that Bayliss only appeared on one of the scientific papers submitted (this was the *Journal of Physiology* paper in 1902, in which Bayliss and Starling first described secretin). The overall subject of Johansson's assessment was the science of hormones, which would now be called "endocrinology." He reviewed the fragmentary knowledge of the pituitary, the thyroid, testes, pancreas, and adrenals. He then outlined how Bayliss and Starling put acid into a denervated segment of small intestine and produced secretion of pancreatic juice. Furthermore, an extract of the mucous membrane of the small intestine could, when injected in to the circulation of an animal, also give rise to pancreatic secretion. The substance released from the mucous membrane had been called "secretin" and Johansson considered it the best characterized hormone.

Johansson goes on to consider two other endocrinological contributions by Starling. The first is a one-off investigation with Janet Lane-Claypon in 1906. (Working in Starling's laboratory in 1904–06, she was the only woman with whom Starling ever published a paper. She appears in the photograph of Pavlov's croquet party in chapter 3.) The authors proposed that during pregnancy in the rabbit, proliferation of the ducts of mammary glands is brought about by a hormone that is released by the fertilized ovum. Lactation begins when production of this substance stops at birth. It is likely that chorionic gonadotrophin is the hormone responsible; Starling and Lane-Claypon deduced its action from their observations. It seems extraordinary that neither author followed up this fundamental discovery. Lane-Claypon left Starling's laboratory shortly after to marry into the aristocracy, becoming Lady Janet Forber.

The third piece of Starling research cited by Johansson is the work with Franklin Knowlton, in which Starling and Knowlton measured the glucose uptake of the heart in the heart–lung preparation. In a diabetic animal, the glucose uptake of the heart was virtually zero; the use of blood to which pancreatic extract had been added increased the uptake of glucose ("reducing sugar") toward normal. (These findings have been discussed earlier).

Johansson also considered the contributions of Schäfer, especially those with George Oliver, in which the relationship between the adrenal glands, adrenaline (epinephrine) and blood pressure were investigated. This is important research, but Johansson asserts that Starling's work is more significant than Schäfer's. Having said all along how outstanding Starling's contributions are, he finally decides not to commit himself: "*It would not hurt to defer somewhat in order to see the development of the ideas presented in all these works*" [emphasis added].

The following year (1914), Johansson heads his "secret document" "*Investigations regarding Langley, Sherrington, Bayliss, Starling, and Einthoven.*" His report contains the same information as the previous year, but rewritten ("The discovery of secretin has been of ground-breaking importance") and with several new names. Once again Bayliss is excluded on the grounds of his only being involved in one publication (Johansson writes as though his 1913 report did not exist). He finished his conclusions with "I regard Starling alone as an eventual candidate for the prize." (One wonders what he meant by "eventual.")

Johansson's thought processes are indeed capricious. For in the same document he suddenly, out of the blue, considers Langley, Sherrington, and Starling for the prize award although his document contains no discussions of Langley or Sherrington's work. This doesn't prevent Johansson from coming to a remarkable conclusion, however: "Sherrington must stand back for Langley and Starling. Which of these two should be placed first, I want to leave undecided for the time being." How does this relate to his previous conclusion regarding Starling alone as an "eventual" candidate for the prize?

No Nobel prizes were awarded in the Great War, and shortly after (in 1920) Johansson recommended the award of a prize to August Krogh, the Swedish physiologist. In retrospect, this seems the death-knell of Starling's hopes, for he was not proposed in the year that Krogh received the prize. Then, in 1926, Johansson returns to his previous form, when Starling is proposed again—this time for his work on the kidney! Johansson, after dismissing this kidney research, briefly reviews Starling's contribution to endocrinology, and says that the work was of such importance that it *should* have been awarded a prize. However, the work had been performed almost a quarter of a century before, and it was his view that awards should be given for recent discoveries. Johansson seems to forget that it was he who put Starling's contribution on the back burner in 1913 and 1914.

So Starling was prevented from receiving a prize by, first, the waywardness of the physiology assessor, who consistently held the view that Starling should have been awarded a prize, but did nothing about it until it was too late, and, second, by the intercession of the Great War.

The whole subject has recently been re-examined by Jens Henriksen of Copenhagen (Henriksen, 2003). Why *did* Johansson do nothing until it was too late? Henriksen proposes that Starling was uncomfortably controversial for the Nobel Committee (and for Johannson) because of his attitude toward Germany. Although he was unambiguous in his feeling against the Germans during the war, he had welcomed German scientists back to the scientific fold after the war. "Moreover" writes Henriksen, "he wrote papers in German journals in the early post-war period, when the general attitude of the British Scientific community and government was to settle harsh terms of punishment on Germany. This conduct was well known in Copenhagen, where, for example, Krogh for patriotic reasons never wrote scientific papers in German . . . this may have contributed to the silence enveloping Starling, in the Nobel Committee during and after the war." The Swedes did not share Starling's forgiving nature. We now have a chain of reasons explaining the non-award of a prize to Starling; if that were not enough, Henriksen provides us with two more observations.

The first is that, over the years, Starling actually received rather few nominations (compared with, say, Sherrington); statistically there is a rough correlation between the number of nominations and the winning of a prize. The second observation relates to Johansson's argument about the too-long delay between the work and the award. Remarkably, as soon as Johansson retired from the committee, several awards were made to scientists whose significant work had been done at least twenty years before; these included Gowland Hopkins and Charles Sherrington. (Starling was no longer alive by this time.)

So everything seemed to conspire to deprive Starling of the award. He was not the first outstanding scientist to be given a poor deal by the Nobel Committee.

The Stockholm Meeting: The "Greatest Game"

The most important physiological event of 1926 was the Twelfth International Physiological Congress, held during August in Stockholm. These congresses occurred (and still do) every three years. Starling went with several dozen British physiologists, and he was accompanied by Florence and Muriel, who were there for the social life. During the congress, a group of physiologists went on a canal trip in a steamer (the *Pallas Athene*) to Saltsjöbadet, and Starling family snapshots show many of the group letting their hair down.

It might be thought difficult to let ones hair down while wearing a wing collar, but Starling managed it. The photo of him conducting some orchestral masterpiece, while smoking a cigarette, sets the tone and provides us with a dust jacket. Another snap shows him cackling ("laughing" seems the wrong verb) along with Cathcart, Lovatt Evans, and Dale. But Starling's aging, which had become progressively more apparent throughout the 1920s, is noticeable in these pictures.

Figure 8-3. Starling on a canal trip (Sweden, 1926) for physiologists attending the International Congress. The figure behind him is probably a physiologist. (*Family collection*)

Figure 8-4. Florence, Ernest, and Muriel Starling on the canal trip in Sweden. The half-person on Florence's right is probably Joseph Barcroft.(*Family collection*)

He made no scientific contribution to the Congress. It included a dinner for 690 guests in the Blue Hall of Stockholm Town Hall, with Starling one of the three after-dinner speakers, the others being Gley from France and von Frey from Germany. Feeling weak, Starling asked Hill if he would deputize for him; Hill refused, saying that it was Starling that everyone wanted to hear.

By all accounts his speech seems to have been memorable, but no transcript of it has survived. His theme was the different ways in which nations pursued physiology, and the English, being a sporting race, saw physiology as a game, and he thought it "the greatest game in the world." We have separate accounts of the speech from Hill and Lovatt Evans: both were moved by it, and remembered the "greatest game" line. In his speech, Starling also proposed that European physiologists should hire a ship to cross the Atlantic for the next international congress, planned for Boston in 1929. The trip (two years after his death) was on the *S.S. Minikadha,* and was a great success.

Descriptions of Starling's public speaking suggest that it gradually improved throughout his life, partly because, according to William Bayliss's son Leonard, he had lessons from a speech coach. By the end of his life he was very good indeed (which presumably contributed to his being asked to speak at the International Congress). Charles Martin wrote: "He had a happy way of finding telling phrases to emphasize the main parts of his discourse, and, when feeling deeply, he was eloquent"(Martin, 1927).

Last Voyage

There was only one more opportunity in Starling's life for him to demonstrate his eloquence. It was in February 1927, when UCL was celebrating its centenary, and he gave the opening address on a century of physiology at the college. It was an affectionate review of his predecessors, of Sharpey, of Burdon Sanderson, of Sharpey-Schafer and Bayliss, with virtually no mention of himself.

Still feeling unwell, he decided that a warm sea voyage would restore his health. A touching belief in the restorative powers of balmy climes runs through his letters, and this time he decided that the Caribbean was going to answer his dreams. He booked a passage with the banana company Elders and Fyffes, who ran return passenger trips to Jamaica; the trips took 34 days, of which 28 were in tropical waters. On April 11, he was seen off by his son John from Avonmouth near Bristol.

The ship was the *S.S. Ariguani*, 1600 tons, carrying 70 passengers and cargo, presumably exchanging the cargo for bananas in the West Indies. All the evidence suggests that Starling was travelling by himself, which is odd for a man who throughout his life was extraordinarily fond of human company. Unfortunately no passenger list has survived, for the Elders and Fyffes records in London were destroyed by German bombs in the 1939–45 war. The *Ariguani* reached Barbados on April 23rd, Trinidad on the 24th and Puerto Limon in Costa Rica on the 28th. The last leg of the outward voyage was four days between Puerto Limon and Kingston, Jamaica. No letters have survived from the voyage, so we know nothing of Starling's state of mind. All that we know for certain is that as the *Ariguani* sailed into the warm waters of Kingston harbor, he was dead.

9

A Life Surveyed

Funeral at Half Way Tree

Extract from the *Daily Gleaner* (Kingston), Wednesday, May 4, 1927:

DR. E. STARLING
MEDICAL EXPERT DIES ON SHIP

Was Professor of Physiology of University College, London, and noted Lecturer and Writer

---

WAS IN POOR HEALTH

Passed away Suddenly when near Port Royal;
Buried at Half Way Tree yesterday.

It is with deep regret we chronicle this morning the death of Dr Ernest Henry Starling, CMG, a noted lecturer and writer on Physiology, which occurred on board the Elders and Fyffes steamer Ariguani on Monday morning shortly before the vessel

reached Port Royal . . . The body was taken to the mortuary at St.
Joseph's Sanatorium; meanwhile, the United Fruit Company
cabled home to the relatives of the deceased for instructions.

Those attending the funeral were: Lieut. Agnew, R.N., ADC
representing His Excellency the Governor, and taking a wreath
sent by His Excellency; Drs. I. W. McLean, C. Strathairn, J.
Geoghegan, Cyril Gideon, R. Howson , K. M. B. Simon, C. A.
Moseley, Ludlow Moody, E. V. Smith, Major Dawson, RAMC,
G. F. Baxter and B. E. Washburn.

The Rev. H. G. Lovell conducted the service and committed
the body to the grave during a downpour of rain. (Starling family
papers)

It was a dismal scenario. A dozen doctors, none of whom could ever have
met Starling, but who might have been brought up on his "Principles of
Physiology," and had a hazy recollection of the Law of the Heart, made up
the damp funeral party.

Dr. McLean, one of these mourners, was medical officer to Elders and
Fyffes, and had signed Starling's death certificate (Chapman, 1962). The
two causes of death that McLean had proposed were "Asthenia [tiredness]
1 year," and "Heart Failure,15 minutes." This was a euphemistic way of saying that he had no idea of the cause of death. The "asthenia" was true, for
Starling had been tired for at least a year—something that McLean had
possibly discovered by talking to the passengers on the *Ariguani*. "Heart Failure" is saying that the heart stopped. We cannot blame McLean for the lack
of evidence, for this could only have been obtained with an autopsy.

If an autopsy had been performed, it is most likely that Starling's body would
have contained secondary tumors, originating from the mass in his colon that
had been removed by Arbuthnot Lane in 1920. The fact that Lane removed
half of the colon suggests that he believed the tumour to be malignant. Presumably no one had ever discussed his prognosis with Starling at the time of
the operation; at that time cancer was not a subject for discussion, not least
because treatment of secondary growths could not be contemplated. Seven
years survival is perhaps fortunate for a patient with such a history.

This is just one of the odd circumstances associated with Starling's death.
Another concerns its date: McLean wrote on the death certificate that the
date of death was May 2—the day that the *Ariguani* docked. But when a
stonemason carved the headstone of the grave, he put April 30 (See "Starling's
grave" in the annotated bibliography for Chapter 9). Without a good reason,
why should anyone want to contradict the death certificate? If Starling had
indeed died on board two days before, it surely would have been noticed. We
are forced to conclude that the headstone date is a strange clerical error, a
rush of blood to the mason's head. It is also curious that no passengers from
the *Ariguani* went to the funeral; they were, after all, the nearest thing to
friends that Starling had. Perhaps the rain was too much for them.

Obituaries at Home

With remarkable speed, a memorial service was held in St. James's, Piccadilly, in London on May 6. The ceremony was, in effect, a funeral, since it enabled his family and friends to mourn. The luminaries of British physiology, including Dale, Sherrington, and Barcroft, were there, as were many staff and students from UCL.

There was a flurry of obituaries. Even to a biographer, naturally biased toward his subject's achievements, their tone seems wonderfully extravagant. Here, for example, is Henry Dale: "All had found in him a generous comrade and leader, and it is impossible to think of physiology in the last thirty years without Starling as the central figure of inspiration . . . he was threatened by physical disaster [his operation] such as, for most men, would have meant the end of all but a restrained activity. Starling's courage was indomitable, his energy and his passion for knowledge flouted all restraint" (Dale, 1927). Starling's pupil Basil Verney wrote:

> He had not only the biological insight to realise what questions were worth asking, but also the knowledge, technical skill, resource and determination for answering them. Starling's personality exemplified the revolt of the human spirit against its "imprisonment in time": his mind shamed composure, and his game enthusiasm infected all with whom he came into contact. He loved physiological science as one would a game . . . Up to within a few weeks of his death, he attracted and was working intensively with, a succession of young men eager to learn from and to catch something of the enthusiasm of one who had become recognised as the greatest living master of experimental physiology. (Verney, 1956)

Charles Martin, Starling's old friend and head of the Lister Institute, wrote: "No-one since Harvey has so greatly advanced our knowledge of the action of the heart" (Martin,1927). It was a golden time for obituaries, for Martin's tribute to Starling in the *British Medical Journal* was over 6000 words long, and overflowing with details of his research. But it avoided delicate topics, such as the missing knighthood, the lack of proper recognition during the war, and the Nobel Prize. Martin could have mentioned his old friend's outspokenness, his enthusiasm for telling it like it was, but he chose not to. Perhaps in 1927 these subjects were not thought suitable for obituaries. In all the tributes there was only one hint of Starling's frankness, and then it was surrounded by more conventional praise: "His pupils and colleagues loved him, and he gave them his best. By nature he was impetuous, but seemed singularly able to inhibit all but the generous impulses, *though he was impatient of fools, officials and rogues*; he loved life and young people, and can be said never to have grown old" (Evans, 1935; emphasis added). Evans, as

Starling's pupil and successor, wrote several obituaries. To modern eyes, the obituaries sometimes seem over the top, verging on pastiche (the world of John Buchan or Bulldog Drummond, perhaps). "The alert and unflinching gaze of blue eyes beneath heavy eyebrows, the smile that so often lit up with extraordinary charm and attraction a face otherwise set in a serious mould by long and arduous study, will endure to the end in the memories of all who knew him" (Evans, 1935). No one writes obituaries like that any more. In fact hardly anyone writes medical obituaries at all; medicine is about today and tomorrow. In Britain, only the *British Medical Journal* continues this strange old custom. If Starling had died in 2003, he would have received half a dozen factual paragraphs in the *BMJ*, with no mention of a face set in a serious mold, or the unflinching gaze of blue eyes.

Florence survived Ernest for less than a year. She died, in January 1928, of cancer. Her estate (£18,028 12s 8d) confirmed the delicate state of the Starling finances, for she had spent much of her life helping lame dogs and supporting good causes, often to the dismay of her family. Daughter Muriel wrote about Florence 23 years after her death: "You will see from the scraps of paper on which she wrote her long letters, one of her minor economies . . . the memories of my mother in my youth are much associated with her economising every penny, and she explained to me that it was so important to her to save in order that Papa should have something to supplement his pension on retirement. And then, after a lifetime of scrimping, he did not live to retire" (Muriel Patterson, 1951).

As we saw earlier, Florence was Ernest's wife, accompanist, deputy, secretary, co-translator, editor, and accountant. Her life might be regarded simply as one of selfless devotion. Yet we have seen on several occasions (such as her attempt to make Ernest give up smoking when he consulted his old chief, Hale-White) what an enterprising person she was. When Starling's Institute was opened in 1909, a great deal of the money for it had actually been raised by Florence, who wrote begging letters to every possible benefactor. We should add "fund-raiser" to the long list of her talents.

In the last chapter of a biography the reader might reasonably expect to be handed some distillate, some essence, of the hero. The problem with this particular hero is that his life was multifaceted: an outstanding scientist who at the same time cared passionately about many worldly matters and was not slow in expressing his views. This outspokenness may well be a UCL trait. Starling, Marie Stopes, J. B. S. Haldane, and A. V. Hill were all UCL scientists with strong "extra-curricular" views, and there are certainly others. (There is a book to be written on the radical thinkers of the college.) Starling was, to use an old-fashioned word, a romantic; he was often swept along by far more overwhelming feelings than the contents of his next scientific paper. And in this he was different from his distinguished contemporaries— physiologists like Sherrington, Dale, Barcroft, or even William Bayliss. These were fine scientists, but they were not men who particularly wanted to change the world. It is worth pointing out that they all received knighthoods, whereas our subject did not.

Three aspects of Starlings life seem to deserve our final reflection. They are: his relationship with Germany, his iconoclastic/educational writings, and his science. The first two overlap to an extraordinary extent.

Starling and Germany

Starling's lifelong affair with Germany began before he qualified in medicine in 1889. He knew by then that he wanted to be a professional physiologist, and encouraged by Leonard Wooldridge, had spent the summer of 1886 in Kuhne's laboratory in Heidelberg. After qualifying in medicine, and marrying Florence, he spent the summer months of 1893 in Heidenhain's laboratory at Breslau. The 1886 visit convinced him that he had to be a physiologist, and the 1893 stay provided him with the research findings (the experiments on lymph) that formed the springboard for his career.

We find Germany wherever we look in Starling's life. Florence came from a German family—the Sievekings. Her father, Sir Edward Sieveking, was physician to Queen Victoria and Albert—a very Teutonic household. Florence had visited Germany many times, where she learned the language and took piano lessons (German music was as admirable as German science). Ernest's sister Gertrude spent two years "finishing" at a school in Hanover.

By the time he was 20, Starling was bilingual in English and German, and by the outbreak of war in 1914, had published six papers in German (out of a total of about 60) There were also translations in Germany of his Arris and Gale lectures on lymph in 1894. This was a great compliment, for there cannot be many occasions in the history of science when results are published simultaneously in two languages. Moreover, his enthusiasm for German medical education itself had become marked. He particularly admired their belief in the scientific basis of medicine, contrasting with the empiricism of the British variety. He believed that Physiology should be taught by people whose lives were devoted to the subject; at Guy's, most of the pre-clinical medicine was taught by amateurs (clinicians) and he found this contrast with German schools painful.

When he came to give evidence to the Haldane Royal Commission in 1910, Starling was uncompromising in his praise for German medical education, a view in tune with Robert Haldane himself, who had a degree from a German university. In giving his extensive evidence, Starling received support from Abraham Flexner and William Osler, a redoubtable duo; the notion of clinical "units," each headed by a professor, became firmly planted in British medicine.

We take the "unit" concept completely for granted now, but if we hark back to the discussions surrounding the Commission, we glimpse the deep suspicion of all things German held by British doctors, and hence the unacceptability of the unit system. The great majority of witnesses were against units, but the Commissioners were evidently not impressed by their reactionary arguments. Starling had nailed his colors to the German mast, and correspondence in contemporary journals hinted at the disapproval that this roused. This was about a year before the Great War began, although we cannot know whether the disapproval was heightened by the possibility of war.

At this time, the "concentration" issue (the concentration of preclinical teaching in two or three medical schools in London) was looming large, and evidence on it was taken by the Haldane Commission. Starling was concentration's most enthusiastic supporter; his motives were based on a desire to increase the efficiency of preclinical teaching in London, but were seen by many as being his way of increasing the political power of UCL. Did the response produced by his pro-German sentiments contribute to the reaction toward concentration? We cannot tell, because the Haldane Commission rejected the idea of concentration, and it took about seventy years for it to be reborn and accepted by London medical schools. (The present situation of five "concentrated" medical schools in London is a *reductio ad absurdum* of what Starling envisaged, for there is no way that 2–400 students a year, doing the same curriculum at a single institution, can receive anything like a university education). It is possible that the concentration pendulum has swung too far, and that the benefits of small-group teaching have been lost.

Because he was a man of passionate beliefs, Starling's pro-German feelings did a complete about-turn when war was declared. He vowed that he would never speak the language again. We have seen how the torpedoing of the *Lusitania* and British hospital ships convinced him of "the absolute evil of the Hun." He clearly had a strong sense of honor—an admirable virtue, but not necessarily a good thing for the subject of a biography, in whom defects of character are usually found more interesting. The end of the war saw him returning to his default position; he was once again an enthusiastic supporter of Germany. Speaking their language, he welcomed German scientists at conferences. In 1923 the International Congress in Edinburgh was ostracized by the French and Belgian representatives because Germans were attending. The conference suddenly became a political football when Sharpey-Schafer and Starling persuaded a group of English physiologists to stay away if the Germans were prevented from attending. This worked, and everyone went to the congress except a few petulant French and Belgian scientists.

In 1924, Starling welcomed a German physiologist, F. Eichholtz, to work at UCL. He was particularly impressed by Eichholtz's previous job description, which was U-boat commander. Rather mischievously, Starling gave him a bench to share with Brull, who had been in the Belgian army. Both men produced publications (on the kidney) while they were at UCL, though there is no account of how they got on with each other. True to himself, Starling resumed publishing papers in German journals after the war (in 1920, 1923, 1924, 1925, and 1926). This may have been a seriously bad career move. We saw in Chapter 8 how the Nobel Prize committee's attitude toward Starling cooled at this time, and Henriksen's suggestion that it seriously damaged his chances of a Nobel Prize. Publishing in German journals was a red rag to the bullish committee. Nowhere in Starling's writings do we find any regret for what he had done, though it is possible that he knew nothing of the committee's attitude toward Germany. But then, given his personality, he might not have been particularly concerned.

Starling's feelings toward Germany were reciprocated, for on his death, obituaries appeared in at least three German journals—something virtually unknown for foreign scientists. L. Asher (1928) placed Starling alongside the most distinguished German physiologists—Ludwig, Weber, and Frank. This was huge praise. M. Von Frey, another obituarist, commented: "Starling was the typical honest, clear-sighted, brave, fair-playing Englishman" ("Starling war der Typhus des aufrechten, klarblickenden, wagemutigen, fair play stets anerkennenden Englanders"). "It is therefore natural that he should participate actively during the war, but when it was over, he re-established in a frank and open-minded way his contact with German colleagues" (Von Frey, 1927). The virtues of this fair-playing Englishman had also made him some new enemies.

The conflict inherent in Starling's relationship with Germany comes from what seems to be a coincidence. Germany, for much of the nineteenth century, led the world in scientific research, and it was this that attracted scientists like Starling. It was then unfortunate that Britain embarked on a lethal war with Germany. Was this actually coincidence? Not quite, for the German talent for science and technology provided the country with the means to go to war with many of its neighbors.

Iconoclast and Reformer

Starling's written views on several subjects led him into controversy, which he rather enjoyed. To use a sporting metaphor of his, he liked being in the scrum. When he was suggesting Hill as the next Jodrell Professor in 1922, he wrote to Mrs. Hill: "I am contemplating, if the Royal Society Council approves, getting out of the scrum and retiring to full-back. I wonder if you know a young and vigorous forward to take my place?" (Mrs. Hill did indeed know such a person.) Hill later commented: "How she replied to his question she does not properly remember; for my wife knows rather little about rugby football" (Hill, 1969). So Starling became full-back and the first Foulerton Research Professor of the Royal Society.

His career in controversy began in 1903, when the *British Medical Journal* published his address to the new medical students at UCL. Described in Chapter 4, it summarized issues that were to occupy him for the rest of his life. London medicine had a complete lack of academic ideals—a formidable obstacle to reform in medical education. Medical organizations were virtually trade guilds, and medical schools "a system of apprenticeships for becoming West End consultants . . . young men, after a struggle, may become West End consultants, but never add anything to the practice of medicine." We have previously seen how he saw the answer in German medical education, where preclinical and clinical departments were headed by professors, capable of applying science to the principles and practice of medicine. Clinical professors had their own beds in hospital, and the patients in these beds were scientifically investigated. Such professors in London, he wrote, would have to be better paid than the current teachers. Occasionally his reforming zeal was a little naïve: "There should be no restriction on private practice . . . since it is quite certain that a man whose life has been spent in the advancement of his subject would not sacrifice his life's work for the love of gain." (When clinical professors finally came into being in Britain, their contracts were very precise about how much private practice was acceptable.) His proposals were not to replace Britain's system of practical, empirical medical education, but to add to it. English medical students, because of the "apprentice" roots of their education, spent a great deal of time with patients on the wards; German clinical students on the other hand, were not welcome in the hospital at all, and had virtually no contact with patients.

The ideas contained in his 1903 lecture provided the basis of his extensive evidence to the Haldane Commission in 1910. His conclusions, especially those relating to the notion of clinical units headed by a professor, overlapped greatly with those of Flexner and Osler (as we have seen.) Starling, Flexner, and Osler, via the Haldane commission, significantly changed the structure of hospital medicine in Britain.

The Great War provided him with a new set of targets for his anger. Now, instead of the dreadful conservatism of the medical establishment, he was confronted by the dreadful conservatism of the army. He was appalled by the inefficiency and lack of scientific knowledge shown by the War Office, and by the Kafka-esque maze of committees that were trying to organize chemical defences. As a major, he wrote a letter to the general in charge of gas: "Is there no-one in charge of this gas business?," earning few points in the club rooms of the War Office.

Toward the end of the war, in September, 1918, after he had resigned his commission, he produced his most spectacular outburst. "Natural Science in Education" appeared in the *Lancet,* and it begins (as noted in Chapter 4): "The astounding and disastrous ignorance of the most elementary scientific facts displayed by members of the government and administration raised doubts in the minds of the British public as to the efficacy of the education imparted to the members of the upper classes." It really seemed like some sort of death wish on Starling's part. He asks how the disastrous ignorance came about, and reviews historically the attitude of the British nation. He finds that "It was always so, except for a brief period in the 1860s when the public stirred in its sleep, under the influence of the publication

of the theory of evolution, and the theological discussions aroused thereby."
He then widens his attack to include the whole English educational system,
with its absurd concentration on classics. ". . . not so much the time that is
spent on the study of Latin or Greek, but the fact that so many years are
passed at school and nothing is learnt. After nine years, nine-tenths of the
boys can read neither Latin nor Greek. They may have acquired a few catch-
words or allusions to classical mythology, but they can give no account of
the way the Greeks lived or the part played by Greek philosophy . . ."

All this he sees as the trademark of the English public school: "Men send
their sons to public schools because there are some advantages which more
than outweigh under present conditions the disadvantages of a senseless
method of education." The advantages consisted of belonging to what Starling calls a "guild" (a favorite word)—a feudal remnant in a society dominated by class. "A boy's success in society will depend more on what he is, or
what guild he belongs to, than on his intellectual equipment. It is indeed
regarded as a heresy to demand of a [government] minister special knowledge of the work which he is appointed to direct, and the idea of promotion by merit in the Army, or other public service, arouses a feeling of horror
in the majority of those belonging to these services." He then develops the
theme of teaching English (rather than classics) thoroughly, with a view to
communicating competently in the language; better still, he argued, to do
this in tandem with a modern language like French.

It is fascinating to compare his conclusions about school education with
those of Abraham Flexner, describing American school children at about
this time: "The American student, unlike the European student, knocks at
nineteen or later for admission at college doors, with his little Latin and
Greek, a bungler in the use of his native tongue, strangely incapable of close
observation or thought and generally without intellectual interests in any
direction" (Bonner, 2002). Starling's views suddenly seem quite restrained.
Starling and Flexner presumably met in 1910 when they both gave evidence
to the Haldane Commission. Flexner wrote later to Starling (in 1922) asking him if he could find a scientific home in England for Pavlov, who was
living perilously in Russia, as we saw in Chapter 7. Fortunately, Pavlov never
had to leave his homeland.

Starling's diatribes include, almost inevitably, attacks on the medical
examination system—he calls it an "incubus." There does not ever seem to
have been a fair way of examining medical students; Starling, heretically,
was proposing continuous assessment rather than short sharp tests of recently
acquired knowledge. "The examination system should therefore be continuous—part and parcel of the education, and not its soulless despot." Here,
he quotes his old hero Thomas Huxley: "examinations, like fire, make good
servants but poor masters" (Starling, 1918). It was at least half a century
before medical educators took continuous assessment seriously.

Carleton Chapman writes memorably: "Starling on education is a grand,
overwhelming phenomenon, vitally expressed and carrying great conviction;

and as a working scientist he is considerably more convincing than many a platitudinous educator . . . It was the heart and his emotions that must have brought him to write about education; but it was his brain, and his contact with the humanitarian intellectuals of his day, that directed his words once he got started . . ." (Chapman, 1962).

Starling's iconoclasm seemed to have little immediate influence. What seems remarkable is that his ideas have invariably turned out to be right, even if this has taken many years. But certain things have not changed: the British are still locked into a class system, and there is still universal ignorance and suspicion of science, which is in some way intertwined with the class system. C. P. Snow's two cultures are alive and well, and Starling would certainly protest loudly about the current science teaching in schools. As we speak, he is in heaven, smoking a cigarette, sitting with Abraham Flexner and arguing about how medical students should be taught.

As Scientist

About forty-three years ago I first made the acquaintance of
Bayliss and Starling . . . Starling, a striking and gallant figure,
once described physiology as "the greatest game in the world";
and his followers still think of it that way. His influence continues
in the Faculty of Medical Sciences which he, more than any
other, helped to create. . . . (Hill, 1962)

It is through his contributions to physiology that Starling will be ultimately remembered, though a historian of London University might say that his greatest achievement was his planning, fund-raising, and building of the Medical Sciences part of UCL. He displayed the same vigor and high intelligence in his schemes for designing his beloved college as he did in his science. In 1920, writing to her daughter Muriel, Florence Starling said "Now I think it would be time for Pups [her rather unlikeable nick-name for her husband] to have a title! If only he could be made dictator of university education in London for 10 years! But we don't think big enough or wide enough . . ." Florence had seen the results in the college of 20 years of Ernest's committee work. (It was not his scientific achievements that were generating her enthusiasm.) She was presumably unaware of the black mark on Pups's file in the office for awarding knighthoods.

To review Starlings scientific achievements we need to return to the early 1890s. This period was a watershed in the history of physiology; the great Carl Ludwig died in 1895 and it marked the beginning of the slide from dominance of German physiology (and a great deal of its other science, perhaps excepting physics). The next fifty or sixty years represented the golden age of British physiology (Eccles, 1971), with Starling's twenty–five productive years right at the beginning. Just how this German/British transition came about has been the subject of a lot of debate; here, we just want to place Starling at the center of the British renaissance His research career began—with William Bayliss—in 1891–2, when the two men recorded the second-ever human electrocardiogram. They followed this by the analysis of peristalsis in the intestine; their research, with that of Walter Cannon, laid the foundations of modern ideas of gut motility, even though their research in this area is rarely quoted.

From then until the end of the century, Starling was busy demonstrating how lymph was formed. As a part of this research, he made the crucial deduction that plasma proteins created an essential osmotic force in the capillary; and that this inward force was similar in magnitude to the outward hydrostatic pressure derived ultimately from the action of the heart. Without plasma proteins, the fluid part of blood would leak rapidly into the tissues. It is a proper token of recognition that the inward and outward forces acting at the capillary have become collectively known as "Starling Forces," and the whole concept known as "Starling's Filtration Principle." The clinical relevance of the work was not appreciated in the 1890s, but its physiological brilliance certainly was. It resulted in his election to the Royal Society and to the Jodrell chair of Physiology at UCL in 1899.

What made this body of work even more remarkable was the "unBritishness" of its subject matter. For body fluids had been up to this time the province of German research, inhabited by Ludwig and his pupils. The remainder of the British physiological renaissance was provided by such men as Gaskell, Langley, and Sherrington at Cambridge, and was mainly concerned with the nervous system. As the renaissance in physiology became more firmly established, other body systems (such as the respiratory and cardiovascular systems) became part of the movement, pioneered by such men as John Haldane at Oxford and Joseph Barcroft at Cambridge. John Haldane was the brother of Viscount Robert Haldane, of the Royal Commission. This book began by berating Oxford and Cambridge for their resistance to science and medicine in the first half of the nineteenth century. Yet by the end of the century they had achieved a total *volte face*, and were as creative and enterprising (in the science of physiology, at least) as anyone in the world.

In 1899, Starling elegantly extended his work on capillary forces to the function of the glomerulus. He anticipated that fluid would only pass across the glomerular capillary if its hydrostatic pressure were greater than the opposing osmotic pressure; urine would only be formed when the perfusion pressure was greater than the osmotic pressure. For some reason this research is seldom quoted, yet was a long way ahead of its time and provided verification of his own (and Ludwig's) ideas on capillary filtration. The reason for the paper's neglect is probably that researchers in the first decade of the century were reluctant to accept that plasma proteins exerted any osmotic effect. The work formed the basis for the work on the kidneys that Starling was to do in the 1920s.

In all this research he had the intellectual and social support of William Bayliss, although his brother-in-law's name only occasionally appears on the title page of the papers. (This may have partly been a consequence of Bayliss's natural modesty.) But it did not stop Bayliss from making practical use of Starling's findings during the war. At that time, blood transfusions for injured soldiers did not exist. The next best thing, according to Starling's findings, would be to provide a solution of plasma proteins to restore blood volume and increase the patient's chance of survival. Such proteins were not available, so Bayliss reasoned that a protein molecule of similar molecular weight—he used gum arabic (beloved by water-color painters)—would work as well. Thousands of injured British troops were given "gum-saline" intravenously, and by all accounts, many lives were saved (Leonard Bayliss, 1961). Gum-saline was an early "plasma expander."

Starling and Bayliss's greatest joint achievement was the discovery of secretin in 1902. The idea of hormones was not entirely new—Schäfer had been writing on "internal secretions" for years, but he failed to focus on the groups of cells ("ductless" or "endocrine" glands) whose role it was to secrete specific messengers into blood. Starling did just this, and produced, in an extraordinarily casual way, the word "hormone" The demonstration of Starling and Bayliss's hormone secretin showed it to have an elegant *modus operandi*: it was made in the wall of the small intestine and released into the blood stream when acid arrived from the stomach. The molecule passed around the circulation and reached the pancreas, which it caused to secrete pancreatic juice. This juice, being alkaline, neutralized any acid present in the intestine. Consequently the release of secretin stopped. This was a mechanism ("a feedback loop") that was later shown to characterize many of the hormones in the body.

The discovery of secretin led to Starling's nomination for the Nobel Prize. We have seen some of the political machinations that accompanied his candidature, and Henriksen has analyzed the issues involved. We have previously discussed an important aspect of this—the Nobel Committee's dismissal of Starling for his German sympathies after the Great War. Starling was never proposed for a Prize for his work on plasma proteins. The research was actually performed in the period 1893–99, and the first-ever Nobel Prize for Physiology or Medicine awarded in 1901.So it would have just been possible for him to be considered for the Prize, if at this time anyone had appreciated the full significance of the work. It actually took 10–15 years for its significance to become apparent. By this time, Johansson would have decided that the research had been done too many years before, and recommended its rejection! It seems to me that both of these subjects—the osmotic pressure of plasma proteins, and hormones—were major contributions to physiology and medicine, and both were worthy of the fluky honor. After establishing his new Institute in 1909, Starling's research centered on the heart. His heart-lung preparation enabled the action of the organ to be studied in isolation, and he showed, with a string of collaborators, that the force of contraction of ventricular muscle was dependent on the resting length of the muscle fibers (his Law of the Heart). In all honesty this wasn't a startlingly new finding. Along the way, he showed that cardiac output was independent of arterial blood pressure (outflow resistance). But what Starling achieved that provided new insight was to integrate the law into the whole circulation—especially in such situations as exercise, and pathological states like heart failure. Starling's Principle and the Law of the Heart both provided paradigm shifts for anyone thinking about the circulation.

Acting as a catalyst, the law has spawned large numbers of physiological and clinical articles. Many of the articles speak of "a Starling curve" as though Starling had provided us with a neat mean curve relating, in man, venous pressure to stroke volume (for example). This idealized curve has a plateau and may or may not have a descending limb, which has become a source of confusion. The confusion is hardly surprising, for Starling produced no such curve. What he did produce, in the third of the four "Law of the Heart" papers (with his son-in-law-to-be, Sydney Patterson) was a graph showing the findings in nine separate dog experiments (Chapter 4). They are an untidy collection of lines. Five of them fail to reach a plateau, and four have a sharp descending limb; this controversial limb seems an idiosyncracy of the heart–lung preparation. It is very difficult to extrapolate Starling's nine lines to the idealized human curve seen in many textbooks. Fortunately for Starling's law, subsequent experiments have shown the general truth of the relationship, with a given heart capable of showing a series (a "family") of parallel curves, the position of the lines capable of changing with circulating adrenaline levels, as may be seen in exercise.

After the war, his heart–lung preparation exerted a remarkable influence over all the research in his Institute. Any physiological problem could be investigated, it seemed, so long as it used the heart–lung preparation. Thus Lovatt Evans used the preparation to examine the fuels used by the heart, Starling and Patterson in their near-miss experiments with insulin, and Starling and Anrep's experiments on the control of blood pressure. This latter research must also be thought of as a near-miss, because Starling and Anrep demonstrated a peripheral mechanism for detecting changes in blood pressure, but somehow failed to localize it to the carotid sinus or the aorta. (Both regions were subsequently shown to be pressure detectors by other investigators.)

When Starling decided to investigate the functions of the kidney, he and Verney naturally used a heart–lung preparation to perfuse the isolated kidney. They made several important observations. The first was that treating the perfused kidney with cyanide produced urine isosmotic with plasma; for tubular activity was destroyed, and the kidney was just acting as a collection of glomeruli. This elegantly supported his neglected 1899 paper on glomerular function. Verney and Starling also noticed that when a normal kidney was perfused, the urine produced by that kidney was more dilute than normal urine though not as dilute as in the cyanide-treated kidney. This was the first

Figure 9-1. The UCL portrait of Starling (not a particularly good likeness) by
W.W. Russell RA, 1926. It actually looks very good hanging in the lecture theatre
of his institute. (*Wellcome Library, London, with permission*)

step in a long trail of research for Verney, a brilliant experimentalist, (who
modestly said that he owed it all to Starling). The research culminated in
Verney isolating a hormone from the brain that could concentrate urine. He
called it antidiuretic hormone, and in its absence diabetes insipidus resulted.

Across the Atlantic, dramatic developments were taking place in kidney
research. These included sampling from individual nephrons (micropuncture)
and the introduction of the concept of "clearance." which revolutionized thinking on the kidney. Starling did not live to witness this renal revolution.

By a strange irony, he was actually proposed for the Nobel Prize for his kidney research. In truth, the work was very competent (and was to lead to Verney's antidiuretic hormone) but it was not quite Nobel Prize standard. Johansson, the Nobel assessor, agreed (Nobel Committee papers, 1926). This seems to have been one of the few rational decisions that he made about Starling's work.

In terms of its physiological significance, Starling's extraordinary research output seems to fall into two phases: from about 1895 to 1914 (when he was 48) he produced his findings on lymph, on the osmotic effects of plasma proteins, on secretin and the idea of hormones, and the Law of the Heart. This latter research (published in October 1914) coincided with the beginning of the war, and he subsequently did no serious research on the heart.

The second phase of his life's research effectively began in 1920, which was the year of his operation. Two episodes of pulmonary embolism after major surgery had a serious effect on his well-being. He was no longer the serious mountaineer that he had been; ballroom dancing and long walks became his exercise. And somehow the edge had gone from his research, for although he had no shortage of talented collaborators, his experiments failed to hit the bull's eye in the way that they had before 1920. Typical of these near-misses were the two projects that involved, respectively, insulin and the control of blood pressure. It is tempting to ascribe his deterioration to spread of the cancer that had been treated in 1920. Of his obituarists, only Henry Dale noted this: "threatened by physical disaster such as, for most men, would have meant the end of all but a restrained activity . . . his energy and passion for knowledge flouted all restraint."

In his range and productivity, in his passionate bridge-building between science and medicine, Starling stood alone. No one person could nowadays achieve the range of subjects that he made his own, and, in his own lifetime, no physiologist approached him in the breadth of his enthusiasms. The man that a group of strangers buried in the rain at Half Way Tree in 1927 was undoubtedly a far more significant figure in human affairs than any of them could have realized.

Appendix I: Starling's Publications

Throughout Starling's scientific life, authors' names in the *Journal of Physiology* papers were in alphabetical order.

1890

Starling EH and Hopkins FG: Note on the urine in a case of phosphorus poisoning. *Guys Hosp Rep,* (3rd series 32)47:275–278, 1890.

1891

Bayliss WM and Starling EH: On the electromotive phenomena of the mammalian heart. *J Physiol,* 12:xx–xxi, 1891 (abstract).

Bayliss WM and Starling EH: Report on the electromotive phenomena of the mammalian heart. *Brit Med J,* 2:186–187, 1891.

Bayliss WM and Starling EH: On the electrical variations of the heart in man. *J Physiol,* 12: lviii–lix, 1891 (abstract).

1892

Bayliss WM and Starling EH: On the electromotive phenomena of the mammalian heart. *Proc Roy Soc B,* 50:211–214, 1892. [Also in *Int. Monatsschrift Anat. Physiol,* 9:256–281, 1892. (Also in English, and slightly longer.)]

Bayliss WM and Starling EH: On some points in the innervation of the mammalian heart. *J Physiol,* 13:407–418, 1892.

Starling EH and Bayliss WM: Note on a form of blood pressure manometer. *Guys Hosp Rep* (3rd series 33), 48:307–310, 1892.

Starling EH: *Elements of Human Physiology.* P. Blakiston, Son, and Co., Philadelphia, 1892. [This work went through 8 editions, the last by J. and A. Churchill, London, in 1907.]

1893

Starling EH: Contributions to the physiology of lymph secretion. (From Breslau) *J Physiol,* 14:131–153, 1893.

Starling EH: Zum Aufsuchen des Peptons in Gewebsflussigkeiten. *Zbl Physiol,* 6: 395–396, 1893.

[Four of Starling's lectures were published in *Guy's Hospital Gazette* (new series, 7) in 1893.They were on lymph, coagulation of the blood, inflammation (and chemotaxis), and diabetes.]

1894

Starling EH and Tubby AH: On the absorption from and secretion into the serous cavities. *J Physiol,* 16:140–155, 1894.

Bayliss WM and Starling EH: Observations on venous pressures and their relationships to capillary pressure. *J Physiol,* 16:159–202, 1894.

Starling EH: The influence of mechanical factors on lymph production. *J Physiol,* 16:224–267, 1894.

Starling EH: Arris and Gale Lectures on the physiology of lymph formation. *Lancet* 1:785–788; 919–923; 990–992, 1894. [Delivered at the Royal College of Surgeons, London, on Feb. 5, 7, and 9, 1894. The series was serialized in German ("Über die Physiologie der Lymphbildung") in *Wien Med Blatter* 17:338–339, 355–356, 371–373, 387–389, 402–405, 419–420, 432–433, 1894.]

Starling EH: On the mode of action of lymphagogues. *J Physiol,* 17:30–47, 1894.

Bayliss WM and Starling EH: On the origin from the spinal cord of the vasoconstrictor nerves of the portal vein. *J Physiol,* 17:120–128, 1894.

Bayliss WM and Starling EH: On the form of the intraventricular and aortic pressure curves obtained by a new method. *Int Monatsschr Anat Physiol,* 11:426–435, 1894.

1895

Leathes JB and Starling EH: On the absorption of salt solution from the pleural cavities. *J Physiol,* 18:106–116, 1895.

Starling EH: On the asserted effect of ligature on the portal lymphatics on the result of intravascular injection of peptone *J Physiol,* 19:15–17, 1895.

1896

Starling EH: On the absorption of fluids from the connective tissue spaces. *J Physiol,* 19:312–326, 1896.

Starling EH: The Arris and Gale Lectures on the physiological factors involved in the causation of dropsy. *Lancet,* 1:1267–1270;1331–1334;1407–1410, 1896, and delivered at the Royal College of Surgeons on Feb.17, 19, and 21, 1896.

1897

Leathes JB and Starling EH: Some experiments on the production of pleural effusion. *J Path Bact,* 4:175–180, 1897.

Starling EH: The Arris and Gale Lectures on some points in the pathology of heart disease. *Lancet,* 1:569–572, 652–655, 723–726. [Delivered at the Royal College of Surgeons, London, Feb 22, 24, and 28, 1897.]

1898

Starling EH: On absorption from the peritoneal cavity. *J Physiol,* 22:xxiv–xxv, 1898 (abstract).

Bayliss WM and Starling EH: The influence of blood supply on the intestinal movements *J Physiol,* 23:xxxiv–xxxv (Fourth International Congress), 1898 (abstract).

Bayliss WM and Starling EH: Preliminary note on the innervation of the small intestine. *J Physiol,* 23:ix–xi, 1898 (abstract).

Starling EH: The production and absorption of lymph (pp. 285–311) and the mechanism of the secretion of urine (pp. 639–661) in *Textbook of Physiology,* ed. Schäfer, EA. First volume, Young J. Pentland, Edinburgh and London, 1898.

1899

Starling EH: A lecture on the electrophysiology of nerve and muscle. *Guy's Hosp Gazette,* new series 13:6–10, 1899.

Bayliss WM and Starling EH: The movements and innervation of the small intestine. *J Physiol,* 24:99–143, 1899.

Starling EH: The glomerular functions of the kidney. *J Physiol,* 24:317–330, 1899.

1900

Bayliss WM and Starling EH: The movements and innervation of the large intestine. *J Physiol,* 26:107–118, 1900.

Starling EH: Four chapters (on respiration, the digestive tract, the urinary tract and the generative apparatus) in *Textbook of Physiology,* by Schäfer EA, second volume, Young J Pentland, Edinburgh and London, 1900.

1901

Bayliss WM and Starling EH: The movements and innervation of the small intestine. *J Physiol,* 26:125–138, 1901. [It seems extraordinary that the *Journal* should publish two papers with identical titles by the same authors; see 1899 above. The two papers actually deal with two different species.]

1902

Bayliss WM and Starling EH: On the causation of the so called "peripheral reflex secretion" of the pancreas. (Preliminary communication) *Proc Roy Soc B,* 69:352–353, 1902. Received Jan 22, 1902.

Starling EH: Oration on the building of the University, foundation day, University College, London, June 5, 1902. London (U.C.L.) Union Society, 1902, 14 pp.

Starling EH. Überblick über den gegenwärtigen Stand der Kenntnisse über die Bewegungen und die Innervation des Verdauungskanals. *Ergebn Physiol,* 1 (Abt.2): 446–465, 1902.

Bayliss W.M. and Starling EH: The mechanism of pancreatic secretion. *J Physiol,* 28:325–353, Sept 12, 1902. (Abstracted in *Zbl Physiol,* 15:682, 1902).

Starling EH: On some pathological aspects of recent work on the pancreas. *Trans Path Soc Lond,* 54:253–258, 1903. Communicated Dec. 16, 1902.

1903

Bayliss WM and Starling EH: On the uniformity of the pancreatic mechanism in vertebrata. *J Physiol,* 29:174–180, 1903.

Starling EH: London's debt to medicine. *Brit Med J,* 2:911–913, Oct. 10, 1903.

Bayliss WM and Starling EH: The proteolytic activities of the pancreatic juice. *J Physiol,* 30:61–83, 1903.

1904

Starling EH: *A Primer of Physiology,* John Murray, London, 1904. 128 pp.

Bayliss WM and Starling EH: The chemical regulation of the secretory process. *Proc Roy Soc B,* 73:310–322, 1904. [Croonian Lecture read March 24, 1904. Abstracted in *Nature,* 70:65–68, 1904.]

Henderson EE and Starling EH: The influence of changes in the intraocular circulation on intraocular pressure. *J Physiol,* 31:305–319, 1904.

Barcroft J and Starling EH: The oxygen exchange of the pancreas. *J Physiol,* 31:491–496, 1904.

1905

Bayliss WM and Starling EH: On the relationship of enterokinase to trypsin. *J Physiol,* 32:129–136, 1905.

Starling EH: On the chemical correlations of the functions of the body. *Lancet 2:* 339–341, 423–425, 501–503, 579–583, 1905. [The Croonian lectures given at The Royal College of Physicians, June 20, 22, 27, and 29, 1905.]

1906

Starling EH*: Mercer's Company Lectures on recent Advances in the Physiology of Digestion:* A. Constable, London, 1906. [Delivered at UCL in 1905.]

Henderson EE and Starling EH: The factors which determine the production of intraocular fluid. *Proc Roy Soc B,* 77:294–310, 1905–6.

Lane-Claypon JE and Starling EH: An experimental enquiry into the factors which determine the growth and development of the mammary glands. *Proc Roy Soc B,* 77:505–522, 1905–6.

Bayliss WM and Starling EH: Die chemische Koordination der Funktionen des Körpers. *Ergebn Physiol* 5 (Abt 2) 664–697, 1906.

1907

Starling EH: The chemical co-ordination of the activities of the body. *Science Progress* 1:557–568, 1906–7. [Translation published in *Zbl Ges Physiol Path Stoffwechs.* Neue Folge 2:161 and 209, 1907.]

Starling EH: Die Chemische Koordination der Körper-tätigkeiten. *Verh GesDeutsch. Naturforsch Aerzte,* Leipzig, 1906:246–260, 1907.

Starling EH: A physiologist's testimony on vivisection. (Abstract) *Brit Med J* 1:573–576, 1907. [This testimony was given by Starling before the Royal Commission on Vivisection; see the minutes of the commission, H. M. Stationery Office 1907.The dates of the testimony were actually Nov. 12, 19, and 20, 1906].

Starling EH: Letter in answer to an anonymous note concerning the London School of Medicine for Women. *Lancet,* 2:1346, 1907.

Starling EH: Campaign letter for election to the Senate of the University of London. *Lancet,* 2:1412–3, 1907. [There is an anonymous anti-Starling letter (by "Justitia") on page 1419 of the same volume.]

Starling EH: Letter concerning Dr Waller's laboratory facilities at South Kensington. *Lancet,* 2:1418, 1907.

1908

Starling EH: *The Chemical Control of the Body,* Harvey Lectures 1907–8. JB Lippincott Co., New York, 1909, pp. 115–131 [Lectures delivered Jan 11, 1908, also published in *JAMA,* 50:835–840, 1908.]

Starling EH: The scientific education of the medical student. *Lancet,* 2:479–480, 1908.

1909

Starling EH: *On the Use of Dogs in Scientific Experiments.* Publications of the Research Defence Society 1908–9. Macmillan and Co., London, 1909, pp. 23–33.

Starling EH: The physiological basis of success. *Science*, 30:389–401, 1909. [Address delivered at Winnipeg, 1909.]

Starling EH: Heart massage in chloroform syncope. *Lancet*, 2:1591, 1909.

Starling EH: *Mercers' Company Lectures on the Fluids of the Body* A. Constable, London, 1909, 186 pp. [Given at UCL in 1907, and also as the Herter Lectures, New York in 1908. These lectures are actually seven as described above, with an extra one that is a repeat of the Arris and Gale lecture, *The Causation of Dropsy* (1896).]

Starling EH: University of London. Description of the new Institute of Physiology at University College. Privately printed, 1909, 16 pp. [This reference is unfindable, but an excellent (anonymous) description of the Institute appeared in *Brit Med J*, 1436–1444, 1909.]

Starling EH: Die Resorption. 1.Die Resorption vom Verdauungskanal aus, in *Handbuch der Biochemie des Menschen und der Tiere*, ed. Oppenheimer, C., Gustav Fischer, Jena.3(2e Hälfte):206–242, 1909.

Kaya R and Starling EH: Note on asphyxia in the spinal animal. *J Physiol*, 39:346–353, 1909.

1910

Bolton C and Starling EH: Note on the blood pressure and lymph flow in a case of heart disease in a dog. *Heart*, 1:292–296, 1910.

Jerusalem E and Starling EH: On the significance of carbon dioxide for the heart beat. *J Physiol*, 40:279–294, 1910.

Starling EH (Obituary in German): Page May. *Ergebn Physiol*, 10:5–7, 1910.

Starling EH: The physiology of digestion, gastric and intestinal. *Guy's Hosp Rep*, 44:141–293, 1910.

1911

[No publications.]

1912

Knowlton FP and Starling EH: On the nature of pancreatic diabetes. Preliminary communication. *Proc Roy Soc B*, 85:218–223, 1912–13.

Knowlton FP and Starling EH: The influence of variations in temperature and blood pressure on the performance of the isolated mammalian heart. *J Physiol*, 44:206–219, 1912.

Knowlton FP and Starling EH: Experiments on the consumption of sugar in normal and diabetic heart. *J Physiol*, 45:146–163, 1912.

Ishikawa H and Starling EH: On a simple form of stromuhr. *J Physiol*, 45:164–169, 1912.

Starling EH: *Principles of Human Physiology*. J. and A. Churchill.1423 pp., 1912. [This large textbook ran to four editions during Starling's lifetime; after his death in 1927, Lovatt Evans took it over. By the eighth edition, it was *Starling's Principles of Physiology*, with Evans as the author. The final edition was in 1956. It was translated into three Spanish editions (in 1925, 1951, and 1955) and one Italian edition (1957).]

1913

Evans CL and Starling EH: The part played by the lungs in the oxidative processes of the body. *J Physiol,* 46:413–434, 1913.

Starling EH: The ethics of antivivisection. *Brit Med J,* 1:950–954, 1913.

Starling EH and Waller AD: letter to the editor (concerning the Nathaniel Alcock Memorial Fund). *Lancet* 2:1430–1431, 1913. Alcock (1871–1913) had taught physiology under Waller at the University of London Laboratories and St. Mary's. In 1911 he became professor at McGill University in Canada, but died shortly after. Starling and Waller established a public fund for Mrs. Alcock and her four children. Alcock must have been a popular figure, because the appeal raised more than £2000 (which was mostly from physiologists).

Markwalder J and Starling EH: A note on some factors which determine the blood-flow through the coronary circulation. *J Physiol,* 47:275–285, 1913.

Murray GR and Starling EH: Discussion on the therapeutic value of hormones. *Proc Roy Soc Med,* 7 (Part 3, Therapeutical and Pharmacological Section):29–31, 1913.

Starling EH: Die Anwendung des Sekretins zur Gewinnung von Pankreassaft, in *Handbuch der biochemischen Arbeitsmethoden,* ed. Aberhalden E. Urban und Schwarzenberg, Berlin. 7:65–73, 1913.

Fuhner H and Starling EH: Experiments on the pulmonary circulation. *J Physiol,* 47:286–304, 1913.

1914

Markwalder J and Starling EH: On the constancy of the systolic output under varying conditions. *J Physiol,* 48:348–356, 1914.

Patterson SW and Starling EH: On the mechanical factors which determine the out put of the ventricles. *J Physiol,* 48:357–379, 1914.

Patterson SW, Piper H, and Starling EH: The regulation of the heart beat. *J Physiol,* 48:465–513, 1914.

Starling EH, Cruikshank EWH, and Patterson SW: The carbohydrate metabolism of the isolated heart–lung preparation. Trans. 17th Internat. Congr. Med. (London) 1913, Sect 2, Physiol. Part 2:83–86, 1914.

Starling EH and Edridge-Green FW: Color vision and color blindness. *Rep Brit Ass Adv Sci,* 1913:258, 1914.

Starling EH and Evans CL: The respiratory exchanges of the heart in the diabetic animal. *J Physiol,* 49:67–68, 1914.

Starling EH, Barcroft J, and Hardy WB: The dissociation of oxyhaemoglobin at high altitudes. *Rep Brit Ass Adv Sci,* 1913:260–262, 1914.

1915

Starling EH: The animal machine and its automatic regulation. *Scientia* 18:185–191, 1915.

1916–1917

[No publications.]

1918

Starling EH: *The Linacre Lecture on the Law of the Heart.* Longmans Green and Co., London, 1918. The Lecture had actually been given at St John's College, Cambridge, in 1915, but the war delayed its publication.

Starling EH: Nature et Traitement du schock chirurgical. *Arch Med Belg,* 71:369–376, 1918.

Starling EH: The significance of fats in the diet. *Brit Med J,* 2:105–107, 1918.

Starling EH: Science in education. *Science Progr,* 13:466–475, 1918–9.

Starling EH: Medical education in England; the overloaded curriculum and the incubus of the examination system. *Brit Med J,* 2:258–259, 1918.

Starling EH: Natural Science in education: notes on the position of natural sciences in the educational system of Great Britain. *Lancet,* 2: 365–368, 1918.

1919

Starling EH: *The Oliver-Sharpey Lectures on the Feeding of Nations; a Study in Applied Physiology.* Longmans, Green and Co., London, 1919. [Delivered to the Royal College of Physicians, June 3rd and 5th, 1919, and abstracted in *Brit Med J,* 1:757–758, 1919.]

Starling EH: Food in relation to health. *Lancet,* 1:591, 1919.

Starling EH: Report on food conditions in Germany. H.M. Stationery office, London, 1919.

1920

Starling EH: On the circulatory changes associated with exercise *J Roy Army Med Corps,* 34:258–272, 1920.

Starling EH: The food supply of Germany during the war. *J Roy Statis Soc,* 83: 225–254, 1920.

Starling EH: *Das Gesetz der Herzarbeit. Übersetzt von Alexander Lipschütz,* E. Bircher, Bonn und Leipzig, 1920.

1921

Starling EH: Heart problems. *Lancet,* 2:1199–1202, 1921. [This was a lecture to the Hunterian Society, Oct. 26, 1921.]

1922

Starling EH: Sur la mechanisme de compensation du coeur. *Presse Méd,* 30:641–645. [This was an exchange lecture given in Paris, 1922.]

Starling EH and Daly I de B: On the effects of changes in intraventricular pressure and filling on ventricular rhythm in partial and complete heart block. *Brit J Exp Path,* 3:1–9, 1922.

Verney EB and Starling EH: On secretion by the isolated kidney. *J Physiol*, 56:353–358, 1922.

1923

Starling EH: The wisdom of the body. *Brit Med J*, 2:685–690, 1923. This was the Harveian Oration given to the Royal College of Physicians, and the lecture was published simultaneously in the *Lancet* 2:865–870, 1923. [It seems strange that the editors of the two journals could not have come to some arrangement.]

Starling EH: *The Action of Alcohol on Man.* Longmans Green and Co, London, 1923, 291 pp. (with Robert Hutchinson and others). [Starling was not particularly interested in this subject but he was asked to lend his name to the book, and in a letter to Muriel says that he was very glad of the money.]

Starling EH: Das Herz-Lungen-Präparat. Ins Deutsche Übertragen von H. Kürten, in *Handbuch d. biol. Arbeits-methoden*, ed by Abderhalden, E., Urban and Schwarzenburg, Berlin, 1923. Abt.V, pp. 827–836.

Starling EH: Hormones. *Nature*, 112:795–798, 1923.

Starling EH: The law of the heart. *Proc Roy Inst Great Britain*, 23:371–376, 1923–4.

1924

Starling EH and Verney EB: Die Folgen der Trennung von Glomerulus und Harnkanälchentätigkeit bei der Säugetierniere. Vorläufige Mitteilung. Translated by Dr R. Plaut. *Pflüger Arch Ges Physiol*, 205:47–50, 1924.

Starling EH: Physiological action of alcohol. *Practitioner*, 113:226–235, 1925.

Starling EH and Verney EB: The secretion of urine as studied on the isolated kidney. *Proc Roy Soc B*, 97:321–363, 1924–5.

Starling EH: The physiology of compensation and decompensation in heart disease. *Cambridge U Med Soc Mag*, 1:249–252, Lent Term, 1924.

1925

Anrep GV and Starling EH: Central and reflex regulation of the circulation. *Proc Roy Soc B*, 97:463–487, 1925.

Eichholtz F and Starling EH: The action of inorganic salts on the secretion of the isolated kidney. *Proc Roy Soc B*, 98:93–113, 1925.

Starling EH and Verney EB: Nachträgliche Bemerkung zu unserer Arbeit Über die Folgen der Trennung von Glomerulus—und Harnkanälchentatigkeit bei der Säugertierniere. *Pfluger Arch Ges Physiol*, 208: 334, 1925.

Starling EH: Ivan Petrovitch Pavlov. Scientific Worthies, xliii. *Nature*, 115:1–3, 1925.

Starling EH: The physiological factors in hyperpiesia. *Brit Med J*, 2:1163–1165, 1925. He gave a second lecture on the same subject in "Discussions on Hypertension," at the annual meeting of the British Medical Association, Bath, 1925.

1926

Starling EH: An improved method of artificial respiration. *J Physiol,* 61:xiv–xv, 1926 (abstract).

Babkin BP and Starling EH: A method for the study of the perfused pancreas. *J Physiol,* 61:245–247, 1926.

Gremels H and Starling EH: On the influence of hydrogen ion concentration and of anoxaemia upon the heart volume. *J Physiol,* 61:297–304, 1926.

Starling EH: William Maddock Bayliss (obituary in German) *Ergebn Physiol,* 25:xx–xxiv.1926. [A similar obituary by Starling appeared in the *Times,* which was unsigned.]

Starling EH: Sir Frederick W. Mott. *J Ment Sci,* 72:317–320, July 1926. [Starling also gave a funeral oration at St. Martins-in-the-fields in June, 1926, and this was abstracted in *Brit Med J,* 1:1065–1066, 1926.]

1927

Starling EH and Visscher MB: The regulation of the energy output of the heart. *J Physiol,* 62:243–261, 1927.

Starling EH: A century of physiology: being the first of a series of centenary addresses, February 28, 1927, University of London, London, 1927, 33 pp. [Abstracted in *Brit Med J,* 1:438, 1927 and *Lancet,* 1:511–512, 1927.]

Starling EH: Why dogs are essential to medical research. *The Fight against Disease* 15:12–14, 1927.

Posthumous Publications

Bayliss LE, Muller EA, and Starling EH: The action of insulin and sugar on the respiratory quotient and metabolism of the heart–lung preparation. *J Physiol* 65:33–47, 1928.

Starling EH: Die Correlation (Integration) der einzel funktionen des Gesamtorganismus, in *Handbuch der normalen und pathologischen Physiologie,* ed Bethe, A and others. Berlin 15(erste Hälfte):1–25, 1930.

Starling's name also appears on collections of reprints that emerged from his laboratory. The first was 1900–1902; the last was 1927, when the *Collected Papers* (volume XXV) was edited by Lovatt Evans, J. C. Drummond, A. V. Hill, and Starling.

He also edited *Monographs on Physiology,* published by Longmans Green, London. These included "The Involuntary Nervous System" (Gaskell), "The Conduction of the Nerve Impulse" (Lucas), "The Vasomotor System" (Bayliss), "The Physiology of Muscular Exercise" (Bainbridge) "The Secretion of the Urine" (Cushny), "Carbohydrate Metabolism and Insulin" (Macleod), "The Internal Secretions of the Ovary" (Parkes) and "The Pressure Pulses in the Cardiovascular System" (Wiggers). After Starling's death, Lovatt Evans and Hill took over the editing of the series.

Appendix II: Publications from the Department of Physiology, UCL, 1899–1927 (Starling's Years)

1899–1909: The decade up to the building of the Institute

The publications are as listed in the Year Book of UCL. SS = Sharpey Scholar; PP = Professor of Physiology; * = Fellow of the Royal Society.

*E.H. Starling (1899–1909) (24 publications)

*W.M. Bayliss (1899–1909) (22 publications)

W.A. Osborne (1899–1902) (7 pubs) SS1901, Ass. Prof. 1902, PP Melbourne 1903 (miscellaneous topics)

T. Swale Vincent (1899–1902) (8 pubs) SS 1902, PP Manitoba (1904–1920), Middx Hosp Med School (1920–1930) (endocrinology)

W. Page May (1900–1909) (11 pubs) Unsalaried. Private lab. Died 1910 (neurology)

J.L. Bunch (1899–1901) (4 pubs) (innervation of bowel)

B.D. de Souza (1901–1911) (2 pubs) (biochemistry)

*F.A. Bainbridge (1900–1906) (6 pubs) PP Durham 1911; St Bartholomew's 1915 (cardiovascular)

C.H. Fagge (1902) (1) Clinician at UCH (innervation of urinary passages)

*J.H. Parsons (1901–1905) (9) SS 1901 Distinguished ophthalmologist.

L.A. Da Silva (1901–1904) (2) (biochemistry)

R.H. Aders Plimmer (1904–1914) (10) (biochemistry). A founder of the Biochemical Society

*S.B. Schryver (1903–1908) (10) (biochemistry)

*H.H. Dale (1901–1906) (5) George Henry Lewes Student 1901–1904; SS 1904. Nobel Laureate

G.A. Buckmaster (1905–1919) (1) Human respiration. Reader in applied Physiology.

J.M. Hamill (1904–1907) (4) SS 1904. Biochemistry (foodstuffs)

Janet Lane Claypon (1904–1907) (4) Published with EHS (Reproduction.)
E.E. Henderson (1904–1906) (3) Ophthalmology: intra-ocular fluids.
F.H. Scott (1907–1908) (3) SS 1907; miscellaneous.
G.C. Mathison (1907–1910) (5) SS 1907–1910; Assistant in physiology; killed
    Gallipoli 1915. EHS wrote his obituary.
*T. Lewis (1907–1909) (1) One of the first Beit Fellows (1910). Cardiology
R. Kaya (1909–1910) (2) Biochemistry
C.C. Lieb (1909) (2) Metabolism
Barbara Ayrton (1909–1910) (2) Biochemistry

2: After the building of the new institute in 1909 (1909–1914)

The authors in the above list published the following number of papers:
Starling (17), Bayliss (7), Aders Plimmer (7), Mathison (4), Page May (2),
    Schryver (2), Thomas Lewis (1), Kaya (1), Barbara Ayrton (1), and
    Buckmaster(1), along with these new names:
Jerusalem (1910) (1) One pub. with EHS; nothing known.
*Bolton (1910) (1) Became pathologist in UCH.
Archibald Smith (1910) (1) Nothing known.
Miss S.C.Eves (1910) (1) Nothing known.
F.W. Edridge Green (1910–1913) (11) Prolific writer on color vision; became
    government adviser on same.
*C. Lovatt Evans (1911–1913) (6) SS 1910 PP Leeds 1918, Barts 1922, UCL 1926.
    Metabolism of muscle.
F. Knowlton (1911–1912) (4) Sabbatical from Syracuse, New York (later PP
    there) Heart–lung prep. Diabetes.
Marjorie Stephenson (1911) (1)
W. Stepp (1911) (1) Preparation of secretin
Ishikawa (1912) (1) Japanese visitor. Worked with EHS on a stromuhr.
*G. Anrep (1912) (2) From Pavlov's laboratory. Two periods at UCL; PP Cairo
    1930. CVS
J. Homan (1912) (1) Histology of pancreas.
S. Hami (1912) (1)
S. Patterson (1913) (2) One of first Beit Fellows; Heart–lung prep with EHS.
    EHS' son-in-law.
J. Markwalder (1913) (2) Heart lung prep with EHS, otherwise nothing known
H. Fühner (1913) (1) German visitor; worked with EHS on pulmonary circulation.

Little or nothing was published from the institute during the Great War.

The UCL Year Book lists 7 publications in 1919 (3 by Starling, 4 by Bayliss.)
    They are all reviews.

3: From 1920–1927 the following were authors

Starling (15), Bayliss (3), Anrep (18), Lovatt Evans (1) and these new names:
*J.C. Drummond (1920–1927) (37) Prof. Biochem. Author: "The Englishman's
    Food" (Nutrition)
D.T. Harris (1920–1922) (4) Asst. Prof.
R. Shoj (1920) (1)
K. Sassa (1920) (1) (CVS)

H. Miyazaki (1920) (1) (CVS)
Katharine Howard (1921–1926) (11) Published with Drummond (Biochemistry)
*I. de Burgh Daly (1920–1924) (3) PP Birmingham 1927 (CVS)
S.S. Zilva (1921) (2)
R.K. Cannan (1921–1927) (5) Published with Anrep (CVS)
H.L. Jameson (1922) (1) Biochemistry
T. Nakagawa (1922) (2) (CVS)
*A.V. Hill (1924–1927) (18) PP UCL 1923; Foulerton Research Professor 1924;
    Nobel Laureate (Muscle)
C.N.H. Long (1925–1926) (6) Published with Hill (Biochemistry)
H. Lupton (1925) (1) Came to UCL with Hill; died unexpectedly 1924 (Biochemistry)
*E.B. Verney (1924–1925) (7) Beit Fellow; Prof Pharmacology UCL 1926. (Renal
    physiology)
*A.S. Parkes (1924–1927) (19); SS (Reproductive physiology)
B. Babkin (1924–1925) (2) Pupil and biographer of Pavlov; PP McGill, Canada
    (Pancreas)
W.O. Fenn (1924) (2) From U.S.; famously worked with Hill (Muscle)
K. Furasawa (1925–1927) (4)
F. Eichholtz (1925–1926) (2) from Heidelberg Worked with EHS and Verney
    (Renal function)
F. Plattner (1925) (1) Probably from Germany (Insulin and heart–lung prep)
H. Gremels (1926–1927) (3) From Marburg (CVS; see Chapter 5)
A.C. Chibnall (1926–1927) (2)
E.W. Cruickshank (1926) (1) (Coronary arteries)
L.N. Katz (1926) (4) From Chicago (CVS)
H.J. Channon (1926–1927) (4) (Biochemistry)
W.K. Slater (1926–1927) (3)
L. Brull (1926) (2) (With EHS and Eichholtz (kidney)
A. Subba Rau (1926) (1) (Coronary arteries)
P. Eggleton (1926–1927) (4) (Biochemistry)
E. Fischer (1926) (1) (Muscle)
H. Goldblatt (1926) (1) (CVS)
R.K. Lambert (1926) (1) (CVS)
Phyllis Kerridge (1926–1927) (2) (Biochemistry of cardiac muscle)
J.R. Pereira (1926) (2) (Metabolism of exercise)
R.J. Lythgoe (1926) (2) SS (Metabolism of exercise)
Vera Reader (1926) (2) (Biochemistry)
*W.S. Duke Elder (1927) (3) Became distinguished ophthalmologist (Vision)
R.W. Gerard (1927) (2) From US (Heat production in nerve)
G.F. Marrian (1927) (2) (Biochemistry)
A. Levin (1927) (1) (Visco-elastic properties of muscle)
J. Wyman (1927) (1) From US; same as Levin
M.B. Visscher (1927) (2) From Minnesota; worked with EHS on metabolism of heart.

This information is mostly derived from the University College Calendar and the Year Book, almanacs that include lists of publications and staff, respectively. The numbers are undoubtedly an underestimate, partly because 1923's publications are totally missing from the Calendar, and partly because it would have been a difficult annual chore for Starling (or his secretary) to collect the titles of the publications from the department. Thus comparison of Starling's official list of publications over these years with those provided by the Calendar gives numbers of 96 and 59, respectively.

The Calendar and Year Books are in The Special Collections Library of UCL.

Annotated Bibliography

Preface

Chapman CB. 1962. Ernest Henry Starling—the clinician's physiologist. *Annals of Internal Medicine* 57(Suppl 2): 1–43.

Colp R. Ernest Starling. 1952. *Scientific American* 185:57–61.

Eccles JC. 1971. British physiology: some highlights, 1870–1940. In: *British Contributions to Medical Science*, Ed. Gibson WC. London: Wellcome Institute for the History of Medicine.

Henriksen JH. 2000. *Ernest Henry Starling (1866–1927). Physician and Physiologist.* Copenhagen: Laegeforeningens forlag.

Starling EH. 1923. The wisdom of the body. *British Medical Journal* 2:685–90.

Prelude

Ashton R. 2000. *G.H. Lewes—An Unconventional Victorian.* London: Pimlico Edition, p. 3. Lewes was one of the founder members of the Physiological Society in 1875, though he subsequently had little to do with the organization. It is likely that Lewes attended medical lectures at UCL, but took no exams (p. 14 in Ashton). "Dilettante" and "polymath" merge imperceptibly in Lewes' character.

Clarke E. 1972. Marshall Hall's entry in *Dictionary of Scientific Biography*. New York: Scribner.

Cope Z. 1966. Private medical schools of London (1746–1914). In: *The Evolution of Medical Education in Britain,* Ed. Poynter FNL. London: Pitman Medical Publishing Co. Ltd.

Desmond A. *Huxley: From Devil's Disciple to Evolution's High Priest.* New York: Penguin Books, 1997. This marvellous biography includes a lot of detail of the social conditions in Victorian England and of a young doctor's attempts to cope.

Eccles JC. 1971. British physiology—some highlights, 1870–1940. In: *British Contributions to Medical Science,* Ed. Gibson WC. London: Wellcome Institute for the History of Medicine.

Harte N and North J. *The world of UCL 1828–1990.* London: UCL, 1991. Many of the subsequent details of the history of UCL are provided by Harte and North.

Huxley T. 1978. On elementary instruction in physiology. In: *Science and Education: Essays.* London: Macmillan, 1893a, pp. 294–302. Huxley was also organizing practical biology courses at the School of Mines in South Kensington. See: Geison GL. *Michael Foster and the Cambridge School of Physiology.* Princeton, NJ: Princeton University Press.

————. The state and the medical profession. In: *Science and Education: Essays* London: MacMillan, 1893b, pp. 323–346. Many of Huxley's essays— especially on medical education—have an extraordinary prescience. He proposes, for example, that preclinical medicine in London should best be taught in two or three large institutions—a suggestion that took about a century to be fulfilled (see Chapter 3).

Physiological Society minute books. The books (1876 onward) are in the library of the Wellcome Institute, London, and provide some details of the early meetings. Contemporary Medical Archives Centre (CMAC): SA/PHY.

Robb-Smith AHT. 1966. Medical Education at Oxford and Cambridge prior to 1850. In: *The Evolution of Medical Education in Britain.* Ed. Poynter FNL. London: Pitman Medical Publishing Co. Ltd.

Sanderson JB. From a copy of the letter in the original minute book of the Physiological Society 1867, Library of the Wellcome Institute, London Contemporary Medical Archives Centre (CMAC): SA/PHY.

Sharpey-Schafer E. *History of the Physiological Society during its first Fifty Years 1876–1926.* London: Oxford University Press, 1927, p. 1. The only account of the personalities of the society, most of whom were known personally by Sharpey-Schafer.

Sykes AH. 2000. Foster & Sharpey's tour of Europe. *Notes and Records of the Royal Society of London* 54(1): 47–52. Michael Foster's memoir, written in 1880, is a revealing account of Sharpey's knowledge of European physiologists. The two made their five-week tour in 1870; the document was written in response to Allen Thompson, Professor of Anatomy in Glasgow, who began to write a biography of Sharpey, but never finished it.

Tansey EM. 1993. In *Women Physiologists,* Eds. Bindman L., Brading A., and Tansey T. London: Portland Press, pp. 3–16. The first formal proposal for women to be admitted to the Physiological Society was made by JS Haldane in Jan 1913. Starling opposed this for a rather original reason. The society, he said, was primarily a dining society, "and it would be improper to dine with ladies smelling of dog—the men smelling of dog that is." Was this genuine concern for women's finer feelings? It hardly matters, for women were formally permitted to join the society in January 1915, when they might even have smelled of dog themselves.

Chapter 1

Anonymous: In an unsigned obituary of Bayliss (it was actually written by Starling) in *The Times*, Aug 28, 1924, we read: "Here in the two rooms [in St. Cuthbert's] assigned to him as a sanctum, one of which served as a laboratory, while the other, lined with books and with a little red lamp constantly burning in front of a picture of the Virgin and Child, provided him with a study . . . " If Bayliss had been a practicing Catholic, he was an unusual one, for he was a great admirer of Marie Stopes and her advocacy of birth control (see Chapter 5).

Barcroft H. 1976. The Bayliss-Starling memorial lecture 1976: Lymph formation by secretion or filtration? *J Physiol* 260:1–20.

Barcroft J. 1928. The entry for "Starling" in: *Dictionary of National Biography*. London: Oxford University Press, 1922–30, pp. 807–809.

Bayliss L. 1960. The details of Bayliss's life come mainly from his son, Leonard (*Perspectives in Biology and Medicine*, 4, 460–478).

Bayliss WM, Starling EH. 1892a. On the electromotive phenomena of the mammalian heart. *International Monatsschrift f. Anat Physiol* 9:256–281.

———. 1892b. On some points in the innervation of the mammalian heart. *J Physiol* 13:407–418.

———. 1894. Observations on venous pressures and their relationship to capillary pressures. *J Physiol* 16:159–202.

Chapman CB. 1962. Ernest Henry Starling—the physician's physiologist. *Ann Int Med* 57 part 2, suppl 2: 1–43.

Frank RG. The tell-tale heart: physiological instruments, graphic methods and clinical hopes, 1854–1914. This outstanding account is in *The Investigative Enterprise*, Eds. Coleman W and Holmes FL. Berkeley: University of California Press, 1988, pp. 230–245.

Heidenhain R. 1891. Versuche und fragen zur lehre von der lymphbilding. *Pflügers Archives*, Bd XLIX.

Kedem O, Katchalsky A. 1958. Thermodynamic analysis of the permeability of biological membranes to non-electrolytes. *Biochem Biophys Acta* 27:229–246. In Kedem and Katchalsky's formulation of Starling's Principle,

$$Jv/A = Lp \{ [P_c - P_i] - \sigma[\pi_p - \pi_i] \},$$

Jv/A is filtration rate per unit capillary area; Lp is wall conductance, P_c and P_i are the pressures of capillary blood and interstitial fluid respectively; π_p and π_i are the two osmotic pressures of plasma proteins in plasma and interstitial fluid respectively. σ is the osmotic reflection co-efficient of the capillary wall to protein. It is always less than 1, because capillary walls are "leaky" to protein.

Levick JR. *Exper Physiol* 76:825–857, 1991; *Phys Soc Mag* 22:26–30, 1996. The modern applications of Starling's Principle have been reviewed by Levick. He examined fourteen sets of experimental data where values of P_c, P_i, π_p, and π_i had been measured, and showed that net filtration pressure was always greater than absorption. The common view that, in an individual capillary, the two somehow balance, is clearly fallacious (except in renal capillaries, where absorption may dominate). Starling himself claimed a balance between filtration and absorption, but he was referring to overall values in the body, not values in individual capillaries. Starling proposed that the hydrostatic/osmotic balance stabilized the circulating blood volume, as we saw in the description of his classic 1896 paper "On the absorption of fluid from the connective tissue spaces."

Martin CJ. 1927. Obituary "EH Starling." *British Medical Journal*, 1:900–905.
Martin knew Starling well, and this obituary is outstanding in its details of his research, but it says little about his life.

Michel CC. 1977. Starling: the formulation of his hypothesis of micro-vascular fluid exchange and its significance after 100 years. *Exper Physiol* 82:1–30.

Pye-Smith PH. 1889. Obituary notice of the late Leonard Charles Wooldridge. *Guy's Hosp Rep* 46:35–42.

Starling EH. 1892. These three letters from Starling are in the Special Collections library in University College London. They are unnumbered.

———. Contribution to the physiology of lymph secretion. *J Physiol* 14:131–153, 1893. Starling's title(s) are given as "Joint Lecturer in Physiology at Guy's Hospital and Grocers' Research Scholar." He supplemented his salary with support form the Grocers' company in the City of London, following a year's support from the British Medical Association. His address is given as "Physiological Institute, University of Breslau."

———. 1894a. The influence of mechanical factors on lymph production. *J Physiol* 16:224–267.

———. 1894b. On the mode of action of lymphagogues. *J Physiol* 17:30–47.

———. 1896. On the absorption of fluids from the connective tissue spaces. *J Physiol*, 19:312–326.

———. Letter to Evans, 17 Jan 1919. In the Wellcome Library (CMAC:PP/CLE)

Starling EH, 1890. Hopkins FG. Notes on the urine in a case of phosphorus poisoning. *Guy's Hosp Rep*, 47:275–278.

Starling EH, Tubby AH. On absorption and secretion into the serous cavities. *J Physiol*, 19:312–326, 1894. Tubby was a Guy's graduate who became a lecturer in physiology there. Afterward he became an orthopaedic surgeon, though, curiously, he did not become a member of the Physiological Society until he had taken up his second career.

Tawara S. *Das Reizleitungssystem des Saugertierherzens.* Jena: Gustav Fischer, 1906. Tawara's name is not the only one associated with the conducting system of the heart. Others include Gaskell (1883), Kent (1893), and His (1893). Keith and Flack (*Lancet* ii Aug 11, 1906) provide a contemporary review.

Chapter 2

Bayliss WM, Starling EH. 1899. The movements and innervation of the small intestine. *J Physiol*, 24;99–143.

Butler S. 1981. Science and the education of doctors in the nineteenth century. PhD dissertation in the University of Manchester.

Cannon WB. 1902. The movements of the intestines studied by means of the Röntgen rays. *Am J Physiol*, 6:251–277.

Geison G. 1978. *Michael Foster and the Cambridge School of Physiology.* Princeton, NJ: Princeton University Press, p. 139. Geison proposes poverty as the main reason for Rutherford's lack of scientific output. Rutherford was *persona non grata* with the Physiological Society for most of the years that he was Professor at Edinburgh. It is not clear why, though differences over vivisection have been suggested.

Guy's Hospital Gazette, May, 1897. In the Wills Library on the Guy's campus.

Guy's Staff Committee and School Committee minute books (1890–99) provide material for this chapter. The books are kept in the Wills library, which was opened as the Medical School Library in 1903, but no longer contains any books. The archives have been moved (2004) to King's College in the Strand.

Harte N, North J. 1991. *The world of UCL 1828–1990.* London: University
College, London.
Krogh A. 1936. *The Anatomy and Physiology of Capillaries.* New Haven, CT: Yale
University Press, p. 283.
Physiological Society minute books, 1897. They are now in the library of the
Wellcome Institute for the History of Medicine (CMAC).
Starling EH. Letter to the Dean, 1895. This letter made such an impression on
the secretary who kept the minute book of the Staff Committee that he stuck
the whole letter in the minutes. (There are no other such insertions in this
large volume.)
———. 1899. The glomerular functions of the kidney. *J Physiol,* 24; 317–330.
Starling F. Florence Starling wrote to her daughter Muriel, on Jan 4, 1922. Her
letter puts the appointment in a slightly different light, for in 1899 Ernest had
also applied for the chair of *pathology* at Cambridge. He was one of two people
on the short list, the winner being Sims Woodhead, from Edinburgh. Florence
wrote: "I remember Langley [the Cambridge Professor of Physiology] writing
'They could all go their own footling ways; he [Starling] should take no further
lot nor pot with them' It stuck in my memory, as it was . . . first time I had
heard the word 'footling' used." Her pretext for this letter was the death of
Sims Woodhead in 1922; at the time, Starling actually contemplated sending
his son-in-law's name for the vacant Cambridge chair (see Chapter 8).
Starling EH, Verney EB. 1924. The secretion of urine as studied in the isolated
kidney. *Proc R Soc B,* 97:321–361.
UCL council minutes, July 28, 1890 (Records office, UCL).
———. July 31, 1890 (Records office, UCL).

Chapter 3

Annual Reports, UCL. Details of the annual changes in UC staff are in the
Annual Report of UCL. Appendix II in the present book includes a list of all
those scientists and clinicians who published one or more papers from the
physiology department, from 1899–1927.
Anonymous annotation. 1889. The pentacle of rejuvenescence. *Brit Med J* 1:1416.
Anonymous. Letter to *Brit Med J* 1418–19. November 16, 1907.
Anonymous. 1909. A new institute of physiology in London. *Brit Med J.* 1:1436–1444.
Babkin BP. *Pavlov—a Biography.* Chicago: University of Chicago Press, 1949, p. 229.
Bayliss WM, Starling EH. The mechanism of pancreatic secretion. *J Physiol*
28:325–353, 1902. The classic paper. It provides most of the background
findings that are reviewed in this chapter, as well as the actual results.
Bernard C. *Mémoire sur le Pancréas.* Paris: Bailliere, 1856 (Translated by J.
Henderson. London: Academic Press, 1985). Bernard showed that tying the
pancreatic duct often produces no pathological changes in the pancreas. This
may be due to the presence of an accessory duct, (present in 50% of dogs)
which enables pancreatic juice to bypass the blockage and enter the duodenum via this alternative route. Or the tied duct may remodel itself around the
ligature, and make a new channel outside the knot, so it is slightly surprising
that Starling expected a ligature of the pancreatic duct to have a significant
effect on the structure of the gland. He may not have known the French
literature as well as he knew the German.
The "Brown Dog" trial is from the *Times* law reports of November 15–18, 1903.
Evans CAL. *The First Bayliss–Starling Lecture: "Reminiscences of Bayliss and Starling."*
Cambridge: Cambridge University Press, 1964. According to Evans, the

bronze statue was given to Battersea Park by a Miss Woodward. Evans implies that Miss Woodward was responsible for the inscription; she sued the Town Council for breach of contract when they had it removed. "That action also went brilliantly in Bayliss's favour, with pointed comments from the judge on the libellous nature of the inscription." (We know nothing of this second case.)

Cushing H. 1940. *The Life of Sir William Osler*. London: Oxford University Press, p. 766.

Franklin KJ. 1953. *Joseph Barcroft*. Oxford: Blackwell, p. 153.

Harte N, North J. 1991. *The World of UCL 1828–1990*. London: UCL (Rev. ed.).

Huxley TH. 1893. "On Medical Education" (1870). This is a reprinted lecture included in *Science and Education*. London: Macmillan and Co.

Martin CJ. 1927. Obituary of EH Starling. *Brit Med J* 1:900–905.

Merrington WR. *University College Hospital Medical School: a History*. London: Heinemann, 1976. The hospital stopped housing patients in the late 1990s. It became known as the Cruciform Building, and started to house the Wolfson Institute for Biomedical Research, and some of the medical school's preclinical facilities. It underwent extensive restoration, recovering lost facades by the use of 200,000 bricks, developed to match the colour and texture of the Victorian originals. It now must look very much as it did in 1906 when it was built.

Needham J. 1936. *Order and Life*. Cambridge: Cambridge University Press, p. 80. Joseph Needham includes this tantalizing footnote: "According to local tradition the word 'hormone' was born in the hall of Caius College, Cambridge. Schäfer or Starling was brought in to dine by [WB] Hardy and the question of nomenclature was raised. WT Vesey, an authority on Pindar, suggested ὁρμάω (excite), and the thing was done." Needham was a fellow of Caius College; his footnote doesn't suggest that he was there at the time. The event is not dated. In another publication, Schäfer gives Starling full credit for introducing "hormone" into the language. This is consistent with it being Starling, not Schäfer, dining with Hardy in Caius College on that particular evening. Schäfer, in fact, gave credit to Starling for the word when he was speaking at an International Medical Congress, in London, reported in the *British Medical Journal*, p. 380, August 16, 1913. Schäfer suggested that "hormone" be reserved for stimulatory agents; he proposed "chalone" for inhibitory substances. Starling was at the meeting, and said that knowledge of hormones was so slight that it was too early to propose such a classification.

O'Connor WJ. 1991. *British Physiologists 1885–1914. A Biographical Dictionary*. Manchester: Manchester University Press.

Pavlov IP. 1967. *Nobel Lectures (Physiology and Medicine) 1901–1921*. Amsterdam: Elsevier, pp. 133–161.

Special Collections Library of UCL. The Pavlov photo is undated, and has only "Northwood" written on its back.

Starling EH. 1903. Introductory Address: London's debt to medicine. *British Medical Journal* 2:911–913.

Webb Beatrice. 1948. *Our Partnership*, Eds. Drake B and Cole M. LSE. Cambridge: Cambridge University Press, p. 99.

Chapter 4

Anrep GV. 1912. On the part played by the suprarenals in the normal vascular reactions of the body. *J Physiol* 45:307–317.

Anrep G. 1956. "Studies in Cardiovascular Regulation," Cooper Lane Medical Lectures. *Stanf Univ Pub Med Sci* 3:199, 1936. See also: JH Gaddum: Gleb Anrep. *Biog Mems Roy Soc* 2:18–34.

Bayliss WM, Starling EH. 1906. Die chemische Ko-ordination der Funktionen des Körpers. *Ergebn Physiol* 5 (Abt 2):664–697.

Burton AC. 1972. *Physiology and Biophysics of the Circulation,* 2nd ed. Chicago: Year Book Medical Publishers, p. 153.

Chapman C, Mitchell J. 1965. *Starling on the Heart.* London: Dawsons of Pall Mall. This anthology of the classical Starling papers has useful commentaries on them all.

Frank O. 1895. Zur dynamik des herzmuskels. *Z Biol* 32:370–447. This was translated by Carleton Chapman and Eugene Wasserman and published in *American Heart Journal* 58:282–317, August, and 467–478, September 1959.

Gregory RA. 1977. The gastro-intestinal hormones. This good historical review is in a book of similar essays: *The Pursuit of Nature,* Eds. Hodgkin AL et al. Cambridge: Cambridge University Press.

Gremels H. 1937. Uber den einfluss von digitalisglykosiden auf die energetischen vorgange am saugetierherzen. *Arch f exper Path u Pharmakol* 186:625–660. (I am grateful to Lady Lise Wilkinson, of the Wellcome Institute for the History of Medicine, for the translations of this and Wezler's work).

Hales S. 1740. *Statical Essays,* 2nd ed. London: W. Innys and others.

Haller, A. 1754. Physiology, being a course of lectures upon the visceral anatomy and vital oeconomy of human bodies. Translated by Samuel Nihles. London: Innys and Richardson, pp. 78–90.

Hamilton WF. 1955. Role of the Starling concept in regulation of the normal circulation. *Physiol Rev* 35:161–168. This is part of a useful symposium on Starling's law: it includes the important work of Sarnoff and Berglund from Sweden showing how in the intact animal (not the heart–lung preparation) a whole family of Starling curves may be constructed with different doses of adrenaline.

Hill AV. IP Pavlov. 1936. *Brit Med J* 1:508–509.

Howell WH, Donaldson F. 1884. Experiments upon the heart of a dog with reference to the maximum volume of blood sent out by the left ventricle. *Phil Trans Roy Soc* 175:139–160.

Katz AM. 2002. Ernest Henry Starling, his predecessors, and the "Law of the Heart." *Circulation* 106:2986.

Knowlton FP, Starling EH. 1912. The influence of variations in temperature and blood-pressure on the performance of the isolated mammalian heart. *J Physiol* 44:206–219.

Krogh A. Letter to Starling. Undated (it was in 1915). Starling had sent Krogh £8, and Krogh replies, pointing out how bad Starling's arithmetic was, and asking for another £5.

Levick JR. 1991. *An Introduction to Cardiovascular Physiology.* London: Butterworths, p. 69.

Markwalder J, Starling EH. 1914. On the constancy of the systolic output under varying conditions. *J Physiol* 48:348–356.

Martin HN. A new method of studying the mammalian heart. *Studies Biol Lab Johns Hopkins Univ* 2:119–130, June 1881–1882.

McMichael J. 1950. *Pharmacology of the Failing Human Heart.* Oxford: Blackwell Publications, p. 11.

Patterson SW, Piper H, Starling EH. 1914. The regulation of the heart beat. *J Physiol* 48: 465–513.

Patterson SW, Starling EH. 1914. On the mechanical factors which determine the output of the ventricles. *J Physiol* 48:357–379.

Roy CS. 1879. On the influences which modify the work of the heart. *J Physiol* 1:452–496.

Starling EH. The Arris and Gale lectures. *The Lancet* I:569–572, February 27; 652–655, March 6; 723–726, March 13, 1897.

———. 1918. *The Linacre Lecture on the Law of the Heart.* London: Longmans Green and Co.

———. 1920. On the circulatory changes associated with exercise. *J Roy Army Med Corps* 34:258–272.

———. 1930. There are several versions of the heart–lung preparation in Starling's publications. This one is from the fifth edition of his *Principles of Human Physiology.*

Wezler K. 1950. Otto Frank zum Gedachtnis. *Zeitschrift f. Biologie* 103(2):96–97.

Interlude

Flexner A. (1866–1959) graduated from Johns Hopkins in 1884 and founded a boys' school in Kentucky. He studied psychology at Harvard and then comparative education in Berlin. In 1908 he published *The American College* and, in 1909, *Medical Education in the United States and Canada,* books that had great influence, and led to Flexner being invited to give evidence to the Haldane Commission. A recent biography of Flexner (*Iconoclast: Abraham Flexner and a Life in Learning* by Thomas Neville Bonner, Johns Hopkins Press, 2002) describes how Flexner and Osler gave evidence to the Royal Commission, but makes no mention of Starling. Starling appears only once in the book (p. 184), where his name is misspelled.

Little EG. *Brit Med J,* (Letter) May 31, 1170–1173, 1913. Little's criticisms of Starling's summary of the Haldane Commission.

Royal Commission, 1913. Its members were: Lord Haldane, Lord Milner, Robert Romer, Robert L.Morant, L. Currie, W.S. McCormick, E.B. Sargent, Louise Creighton; John Kemp, H. Frank Heath (Secretaries). The summary is in volume XL of the Royal Commissions Reports (1913). The reports (there are four volumes that include, inter alia, the Haldane report) are in the library of the University of London in Senate House. The parts of the commission are distributed in 1910 (XXIII), 1911 (XX), 1912 (XXII) and this summary, in volume XL. (The strange order of these Roman numerals is correct.)

Starling EH. We have seen an account of his radical lecture to UCL students (Chapter 3) in 1903, where many of these views first appeared in print.

———. 1913a. The report of the Royal Commission in reference to medical teaching in London. *Brit Med J,* May 17, 1063–1066.

———. 1913b. The report of the Royal Commission in reference to medical teaching in London. *Brit Med J,* May 31, 1168–1172.

Chapter 5

Annual Reports UCL. Details of the wartime lectures in these reports (Now in the Special Collections Library, UCL).

Bell, Hills, and Lucas. Leaflets in the Wellcome Library for the History of Medicine: CMAC (RAMC box 156/751).

Briant K. 1962. *Marie Stopes. A biography.* London: Hogarth Press, pp. 91–92, 132.

Drummond JC, Wilbraham A. 1939a. *The Englishman's Food.* London: First Edition. London: Jonathan Cape, p. 520. Jack Drummond (1891–1952) was Professor of Biochemistry at UCL, and was one of the diners welcoming

Starling back to the department after his operation in 1920. He was murdered on holiday in France (the murder was never solved).

Drummond JC, Wilbraham A. 1939b. *The Englishman's Food*. London: First Edition, London: Jonathan Cape, p. 522.

Evans CL. Reminiscences of Bayliss and Starling. *Journal of Physiology* 1964. (Lecture given 22 March, 1963).

Evans CL. Letters to EH Starling. May 20, 1916. Wellcome Library for the History of Medicine (CMAC. PP/CLE)

Haber LF. 1986. *The Poisonous Cloud. Chemical Warfare in the First World War*. Oxford: Clarendon Press. Haber makes little or no reference to the gas attack by Austria on Italy in June 1916 in which 4,000 Italian troops are reported killed, and which led to Starling's tour of the Italian gas centers described in this chapter.

Hale-White's obituary of Starling. 1927. *British Medical Journal* 1:941.

Krogh A. 1915. Letter quoted in Schmidt-Nielsen B. *August and Marie Krogh. Lives in Science*. American Physiological Society. New York: Oxford University Press, 1995, pp. 109–128.

Major Mackilly (Chemical adviser, British Forces in Italy) wrote his report on Dec 8, 1918, in which he describes this enormous number of SBR's to be sent from England (Public Record Office. PRO.WD 431).

Minutes of the Royal Society Food (War) Committee in the Royal Society Library, London (CMB 72/73).

Minutes of the Royal Society Food (War) Committee during 1917 in the Royal Society Library, London (CMB 72/73).

Minutes of the Royal Society Food (War) Committee in the Royal Society Library, London (CMB 72/73). WB Hardy to Lord Devenport's deputy (undated).

Minutes of the Royal Society Food (War) Committee in the Royal Society Library, London (CMB 72/73). Unsigned memorandum, Dec 22, 1917.

Minutes of the Royal Society Food (War) Committee in the Royal Society Library, London (CMB 72/73). Hardy's sharp letter to Starling (Jan 9, 1918) is included in the minutes.

Palmer A. 1999. *The Gardeners of Salonika*. London: Andre Deutsch, 1965. A very readable account of the war on this front, emphasizing its complexity, for each country involved had a different motive for fighting in this region. (Another useful source is Misha Glenny, *The Balkans*. London: Granta Books).

Starling EH. 2000. From a letter to A. Krogh in the Royal Library, Copenhagen. *August Krogh Archives. Tilg.* 459, 1914. Quoted by Henriksen, J. *Ernest Henry Starling (1866–1927)*. Copenhagen: Laegeforeningens forlag.

———. 1915a. Letter quoted in Schmidt-Nielsen B. *August and Marie Krogh. Lives in Science*. American Physiological Society. New York: Oxford University Press, 1995.

———. to Colonel Horrocks, Nov 7, 1915b. Wellcome Library for the History of Medicine. CMAC. PP/CLE.

———. to Colonel Horrocks, Feb 22, 1916a. Wellcome Library for the History of Medicine, CMAC. PP/CLE.

———. to General Atkins, May 26, 1916b. Wellcome Library for the History of Medicine. CMAC. PP/CLE.

———. to Merringham (?), May 29, 1916c. Wellcome Library for the History of Medicine. CMAC. PP/CLE.

———. letter to his mother Nov 9, 1916d. (Family collection).

———. Nov 27, 1916e.

———. Jan 3,1917a.

———. March 29, 1917b.

———. June 3, 1917c.

———. July 11, 1917d.

Starling's report on Italian gas June–July, 1917e. He found some good things and some bad things. Haber (q.v.) says that Starling's report dismissed Italian facilities, which is not quite true (Public Records Office WD431).

Starling EH. 1919. *The Feeding of Nations.* The Oliver Sharpey lectures. London: Longmans Green and Co.

———. 1920. The food supply of Germany during the war. *Journal of the Royal Statistical Society* 83:225–245.

Chapter 6

Apart from the three letters in the Wellcome Institute, quoted below, much of this chapter is derived from Starling family letters.

Annual Report, UCL, 1923–24. In the Special Collections Library of UCL.

British Medical Journal, Jan 3, 1920, accompanied by announcements of the medical and surgical unit professors in London.

Dale HH. Letter Jan 20, 1919a. Wellcome Institute Library (CMAC:PP/CLE)

———. Letter Jan 22, 1919b. Wellcome Institute Library (CMAC:PP/CLE)

Daly, I de B. 1967. The Second Bayliss-Starling Memorial Lecture. Some Aspects of Their Separate and Combined Research Interests. *J Physiol,* 191, 1–23.

Evans CAL. 1964. *The First Bayliss-Starling Memorial Lecture. Reminiscences of Bayliss and Starling.* Cambridge: Cambridge University Press, 1–17.

Merrington WR. 1976. *University College Hospital Medical School: A History.* London: Heinemann.

Schäfer EA. 1916. Science and classics in modern education. *Nature* 97 (May,18):251–252.

Starling EH. 1918. Natural Science in education. *Lancet* 2:365–368.

———. Letter to CAL Evans, Jan 17, 1919; Wellcome Institute Library (CMAC: PP/CLE).

Chapter 7

Babkin BP. 1949. *Pavlov—a Biography.* Chicago: University of Chicago Press. Chapter 17, Pavlov and the Bolsheviks, includes this part of his life.

Bainbridge FA, 1914. Evans CAL. The heart lung kidney preparation. *J Physiol* 48:278–286.

Banting FG, 1922. Best GH. The internal secretion of the pancreas *J Lab Clin Med,* 7:251–66.

Barcroft H. 1976. Bayliss-Starling Memorial Lecture: Lymph formation by secretion or filtration? *J Physiol* 260:1–20. Henry Barcroft (Joseph Barcroft's son) began his lecture: "I must be one of those few here tonight to have met Starling."

Barcroft J (Joseph). 1924. W Bayliss in *Dictionary of National Biography.* London: Oxford University Press, pp. 69–70. In the last sentence, Barcroft actually wrote "intercourse," not "conversation" but this seemed rather an unfortunate note on which to end our references to Bayliss's life.

Bayliss WM. 1915. *Principles of General Physiology.* London: Longmans Green and Co.

Cannon WB. 1932. *The Wisdom of the Body.* New York: W. W. Norton.

Cushny AR. 1917. *The Secretion of Urine.* London: Longmans Green and Co.

Daly I de B. 1967. Bayliss-Starling Memorial Lecture: Some aspects of their separate and combined research interests. *J Physiol* 191:1–23.

Daly I de B, Pickford LM. Ernest Basil Verney (1894–1967). *Biograph Mem Roy Soc* 16:523–542, 1968.

Drummond J, Wilbraham A. 1939. *The Englishman's Food.* London: Jonathan Cape.

Flexner A. 1921. In The Royal Society library under "Bayliss" and "Starling" correspondence.

Gillies H. 1956. The navel on the knee. *The Practitioner* 177:512. He gives his patient's name in this case report, which is extraordinary. Starling's grandson, Tom Patterson, a plastic surgeon, was Gillies' houseman in the 1940s.

Henderson JR. 2000. Mrs. Starling's leg. *J Roy Soc Med* 93:384–86.

Hill AV. "The Present Tendencies and the Future Compass of Physiological Science." Inaugural lecture at UCL Oct 16, 1923. In: *The Ethical Dilemma of Science and Other Writings.* London: Scientific Book Guild, 1962, p. 8.

Koestler A. 1971. *The Case of the Midwife Toad.* London: Pan Books.

Pavlov IP. 1927. *Conditioned Reflexes.* Translated by G. Anrep. Oxford: Oxford University Press. (Reprinted by Dover Books, 1960.)

Royal Society: Foulerton Research Fund minutes. Nov. 22,1922. CMB 64. LXXIII.

Royal Society Records, 1992, p. 80. CMB 64.

Smith HW. 1959. *From Fish to Philosopher.* New York: Doubleday, pp. 31–48.

Starling EH. 1899. The glomerular function of the kidney. *J Physiol* 24:317–330. A paper that is extraordinarily neglected in the physiological literature; the chapter on glomerular filtration in *The Handbook of Physiology* (ed. E. E. Windhager, Washington, D.C., American Physiological Society: 1992) has 673 references on the subject without mention of Starling's work.

———. Letter to Muriel Patterson from her father (at Taviton Street) Nov. 10, 1922.

———. 1923. The wisdom of the body. *Brit. Med. J* 2:685–90. (identical lecture in *Lancet* 2: 865–70, 1923.)

———. 1923–24. The Law of the Heart. *Proc Roy Inst Great Britain* 23:371–376.

———. 1925. Scientific Worthies XLIII: Ivan Petrovitch Pavlov. *Nature* 115:1–3.

Starling EH, Verney EB. 1924–25. Secretion of urine as studied on the isolated kidney. *Proc Roy Soc* 97B:321–363. A very detailed paper, full of new observations; it was basically the work described here that led to Starling's second nomination for the Nobel Prize (see Chapter 8).

Todes D. 2002. *Pavlov's Physiology Factory.* Baltimore, MD: Johns Hopkins University Press, p. 142.

Wearn JT, Richards AN. 1924. Observations on the composition of glomerular urine. *Am J Physiol* 71:209–227.

Chapter 8

Anrep GV, Starling EH. 1925. Central and reflex regulation of the circulation. *Proc Roy Soc B* 97:463–487.

Banting FG, Best CH. 1922. The internal secretion of the pancreas. *J Lab Clin Med.* 7:251–266.

Chapman C. Ernest Henry Starling. 1962. *Annals Int Med* 57 (Suppl 2). p. 42.

Daly I de Burgh: C. Lovatt Evans. 1970. *Biograph Mem Roy Soc* 16:233–252.

Evans CAL. 1964. The First Bayliss-Starling Memorial Lecture. *Reminiscences of Bayliss and Starling*. Cambridge: Cambridge University Press, 1–17.

Henriksen JH. 2003. Starling, his contemporaries and the Nobel Prize. *Scand J Clin Lab Invest* 63:1–64.

Hering HE. 1927. *Die Karottissinusreflexe auf Herz und Gefässe*. Leipzig: Steinkopff.

Heymans J-F, Heymans C. 1926. Recherches physiologiques et pharmacodynamiques sur la tête isolée du chien. *Arch Int Pharmacodynam* 32:1–33.

Huxley, AF. Letter from Sir Andrew Huxley to the author, May 2003.

Knowlton FP, Starling EH. 1912a. The influence of variations in temperature and blood pressure on the performance of the isolated mammalian heart. *J Physiol* 44:206–219.

———. 1912b. Experiments on the consumption of sugar in normal and diabetic heart. *J Physiol* 45:146–163.

Ludwig C, Cyon E. 1866. Die Reflexe eines der sensiblen Nerven der Herzen auf die motorischen Nerven der Blutgefässe. *Ber Sächs Ges (Akad) Wiss* 18:307–329.

Martin CJ. 1927. Obituary of EH Starling. *Brit Med J.* 1:900–905.

Nobel Committee (at the Karolinska Institute). Reports dated 1914 and 1925–6. They were kindly translated for me from Swedish by Dr. Inga Palmlund.

O'Connor WJ. 1991. *British Physiologists 1885–1914*. Manchester: Manchester University Press.

Schaepdryver AF. 1973. *Corneille Heymans. A collective biography*. Ghent, Belgium: The Heymans Foundation. This contains an excellent review of Heyman's work on baro- and chemoreceptor reflexes by Eric Neil, Professor of Physiology at the Middlesex Hospital Medical School.

Starling, EH. This technical letter to Muriel was not wasted on her, for she had a degree in physiology. She actually had been a student at University College (with a first class degree) though there is no reference in any letter to her attending her father's department. She did not work as a physiologist, but became a civil servant until her marriage to Sydney Patterson.

Tepperman J. 1988. The discovery of insulin. In: *Endocrinology—People and Ideas*, Ed. McCann SM. Bethesda, MD: American Physiological Society, p. 290.

Chapter 9

Asher L. 1928. Ernest H. Starling (Obituary) *Ergebn. Physiol* 27: XV.

Bayliss L. 1961. William Bayliss, 1860–1924: Life and scientific work. *Persp Biol Med* 4:460–447. I have been able to find only the vaguest descriptions of the efficacy of gum-saline on the battle field. It was presumably superseded by blood transfusion after the Great War.

Bonner TN. 2002. *Iconoclast. Abraham Flexner and a Life in Learning*. Baltimore, MD: Johns Hopkins Press, p. 43.

Chapman, C. 1962. A copy of the death certificate is in Carleton Chapman's "E. H. Starling—the clinicians physiologist." *Ann Int Med* 57, suppl. 2, p. 17. The date of death on the certificate is given as May 2; the date of signing is May 5. The ship's name is given as the *SS Origani* [sic]. In the column under "Signature, Qualifications, and Residence of Informant" we read: "I. W. McLean who caused the body to be buried. United Fruit Company, Harbour Street, Kingston." This company was the agent for Elders and Fyffes, Ltd. McLean gave his qualifications as "MD, Maryland, USA."

Dale HH. 1927. (Tribute to Starling) *Brit Med J* 1:905.

Eccles JC. 1971. British physiology—some highlights, 1870–1940. In: *British Contributions to Medical Science*, Ed. Gibson WC. London: Wellcome Institute for the History of Medicine.

Evans CL. 1935. British masters of medicine. Ernest Henry Starling (1866–1927). *Med Press and Circular* London 190:536–541.

Hill AV. 1962. Science and witchcraft, or the nature of a university. In: *The Ethical Dilemma of Science*. London: The Scientific Book Guild, p. 91.

Hill AV. 1969. The third Bayliss-Starling memorial lecture. Bayliss and Starling and the happy fellowship of physiologists. *J Physiol* 204:1–13.

Martin CJ. 1927. Obituary of E.H. Starling. *Brit Med J* 1:900–905.

Nobel Papers. The Nobel Committee at the Karolinska Institute (Postal address: Box 270, SE – 171 77 Stockholm) provided the reports in Swedish. Dr. Inga Palmlund kindly translated them into English.

Patterson Muriel. This letter (October 26, 1951) was written to Ralph Colp (the American physician who was planning a biography of Starling at the time). It is full of affectionate memories of her father "He and I did have many good times together. He took me to my first hearing of the Ring at Covent Garden, having subscribed to two gallery seats (the cheapest, with backless benches to sit on) for the whole cycle." Colp dutifully returned the letter to Muriel, for it was among the family papers provided by Tom Patterson (Muriel's son).

Starling EH. 1918. Medical education in England: the overloaded curriculum and the incubus of the examination system. *Brit Med J* 2:258–259.

Starling family papers. This is from a typed copy of the newspaper report.

Starling's grave. Starling's grave lay almost untouched for about 40 years, although a Professor of Physiology in the University of West Indies, Ian McKay, knew its whereabouts and visited it during the 1960s and early 1970s. Interest in the grave suddenly increased in 1974, when McKay—by now retired—helped various external examiners from the U.K. to identify the headstone in the dilapidated cemetery. Among these examiners were Sidney Hilton (Senior secretary of the Physiological Society and professor at Birmingham) and George Brownlee (Professor of Pharmacology at King's College, London). It was agreed that funds should be provided, half from the Physiological Society and half from UCL. The funds were to be used for the upkeep of the grave and for a new Starling memorial plaque on the campus.

In October 1977, Hilton visited the physiology department as external examiner. The year was the 50th anniversary of Starling's death, and Hilton unveiled a marble plaque (costing about £150) on the lawn outside the physiology department. He sent a photograph of the plaque to Denis Noble, his successor as Senior Secretary of the Physiological Society.

Every so often the Secretary receives a letter from a visitor to Jamaica: does the society know that Ernest Starling is buried there?

The correspondence relating to the grave is in the CMAC collection of the Wellcome Institute (SA/PHY/E.3/12)

Verney EB. 1956. Some aspects of the work of Ernest Henry Starling. *Ann Sci* 12:30–47.

Von Frey M. 1927. Ernest Henry Starling (Obituary). *Munchen Med Wochenschr* 74:898–899.

Index

215

Printed and bound by CPI Group (UK) Ltd, Croydon, CR0 4YY

03/10/2024

01040425-0002